THE ECONOMICS OF HIGHWAY PLANNING

CANADIAN STUDIES IN ECONOMICS

A series of studies, formerly edited by V. W. Bladen and now edited by Wm. C. Hood, sponsored by the Social Science Research Council of Canada, and published with financial assistance from the Canada Council.

1. *International Cycles and Canada's Balance of Payments, 1921-33.* By Vernon W. Malach.
2. *Capital Formation in Canada, 1896-1930.* By Kenneth Buckley.
3. *Natural Resources: The Economics of Conservation.* By Anthony Scott.
4. *The Canadian Nickel Industry.* By O. W. Main.
5. *Bank of Canada Operations, 1935-54.* By E. P. Neufeld.
6. *State Intervention and Assistance in Collective Bargaining: The Canadian Experience, 1943-1954.* By H. A. Logan.
7. *The Agricultural Implement Industry in Canada: A Study of Competition.* By W. G. Phillips.
8. *Monetary and Fiscal Thought and Policy in Canada, 1919-1939.* By Irving Brecher.
9. *Customs Administration in Canada.* By Gordon Blake.
10. *Inventories and the Business Cycle with Special Reference to Canada.* By Clarence L. Barber.
11. *The Canadian Economy in the Great Depression.* By A. E. Safarian.
12. *Britain's Export Trade with Canada.* By G. L. Reuber.
13. *The Canadian Dollar, 1948-58.* By Paul Wonnacott.
14. *The Employment Forecast Survey.* By Douglas G. Hartle.
15. *The Demand for Canadian Imports, 1926-1955.* By Murray C. Kemp.
16. *The Economics of Highway Planning.* By David M. Winch

THE ECONOMICS OF HIGHWAY PLANNING

David M. Winch

UNIVERSITY OF TORONTO PRESS

Printed in the Netherlands

PREFACE

THE AUTHOR wishes to express his gratitude to all those who have helped by their support, encouragement, assistance and criticism, both in the writing of the original study, which was accepted in 1957 for the degree of Doctor of Philosophy in the University of London, and in the task of revision for publication. The list would be too long to mention all by name, but the opportunity cannot pass without recording a word of thanks to those without whom his book would never have been finished, nor even begun: to the four bodies who made the research possible, the University of London for the Gerstenberg Studentship, the Royal Insurance Company for a Fellowship at Yale University, the Trustees of the Rees Jeffreys Road Fund for a studentship at the London School of Economics, and the Canada Council for a grant to assist in revision for publication; to my three supervisors while a graduate student under whose guidance the research was done, Mr. G. J. Ponsonby and Mr. R. Turvey of the London School of Economics, and Professor K. T. Healy of Yale University; to Miss Anne Tyler for typing the manuscript; and to the University of Toronto Press and the Social Science Research Council of Canada for arranging publication.

A special word of thanks is due to the many officials in Washington, D.C., State and Provincial Highway Departments in the U.S.A. and Canada and the Road Research Laboratory in England who gave so generously of their time to discuss these problems with me. Many of those most intimately concerned with economic problems expressed their need for some guidance from an economist. Their encouragement has sustained me through the scepticism of others. My hope is that they will now try in practice the approach advocated here. Over a hundred years ago Jules Dupuit, engineer and economist, wrote,

> Les fonctions de l'ingénieur des Ponts et Chaussées touchent à trop de points de l'économie politique pour que cette science soit demeurée étrangère à nos études. L'usage en a fait une science morale: le temps en fera, nous en sommes convaincu, une science exacte qui, empruntant à l'analyse et à la géometrie leur procédés de raisonnement, donnera à ses démonstrations la précision qui leur manque aujourd'hui.

His hopes have not yet been fulfilled and in recent decades the two professions have gone their separate ways. If this book helps them to work more closely together on problems of common interest it will have served its purpose.

D. M. WINCH

University of Alberta
Edmonton, Canada
July, 1962

CONTENTS

INTRODUCTION

THE GROWTH in road transport in recent years has focused attention on the importance of a well-planned highway system to the economy of any nation. So complex are the problems involved in achieving an optimum highway system, however, that three different professions are involved, politics, engineering, and economics. Highways are built by society for the use of society and primary responsibility therefore rests with governments. But the days when local councils simply decided to put more gravel on a road and hired unskilled labour to spread it are long since past. The design and construction of modern highways is a complicated technical business and the politicans have therefore turned to the engineer for advice. The profession of highway engineering has grown up, and as the administrative tasks of organizing highway departments and planning highway development arose it was natural that the engineer should add these responsibilities to his field. Today the whole business of highway planning and construction still rests very largely on the teamwork of politican and engineer.

The impact of highways on the economy as a whole is so great, and the philosophical problems involved in a society collectively making decisions about the proper extent of use of its resources for highways so intricate, however, that the economist also has a role to play. But the economics of highway planning has not grown as a single branch of the subject. Instead it has developed from two opposite ends, and failed to meet in the middle. The academic economist has toyed with highways along with other examples of public enterprise as problems to which the simple economics of private enterprise do not apply. The whole field of welfare economics has grown up as a result of efforts to find answers to the problems of what constitutes optimum economic action for a society collectively. But the more the welfare economist searches for precision the more elusive it seems, while the literature on the subject becomes ever more complicated and incomprehensible to anyone but the specialist. The professional economist has therefore had very little impact on the practical world of highway planning. He will not offer advice because he cannot be sure of his answers, while the administrator who is faced with the problems cannot get much guidance from the specialist literature. But the administrator, an engineer by profession, has to find answers to the economic problems with which he is faced and cannot afford to wait until welfare theory is perfected before making a decision. He has instead applied the tools of accounting and arithmetic and developed techniques such as the cost benefit analysis, which do at least give answers, though they might not be the right answers. To such techniques he has given the name "highway economics."

Thus we have reached the stage where economic theory can fulfil a very useful role in offering sound guidance, albeit not perfection, on the problems of an economic nature which are involved in highway planning; yet this knowledge is not used. Instead we turn to the engineer for solutions not only to those technical problems on which he is an expert, but also for advice or decisions on matters of economics about which the engineer is essentially a layman. The resourcefulness of the engineer is attested by the quality of the highways which we have today, and it is as well that if only one profession were to be involved we turned to the engineer for advice on economics rather than to the economist for advice on engineering. But the fact remains that decisions as to how many highways are worth their cost, how many lanes it is worth building, what is the best location for a highway, how should we raise the money to pay for the highways, and what effects will alternative systems of financing have on the transport industry and the economy as a whole—all these decisions are concerned with economics, not engineering, and it is to the economist that we should look for guidance, not to the engineer.

This book attempts to extract from economic theory the best guidance available on these problems, and to present it in a form which can be understood by the non-economist. The author is not an engineer and such data of an engineering nature as are cited are for illustrative purposes only. The author craves the indulgence of the professional engineer if these are not the best or most recent data. The prime object of the book is to present the techniques of economic analysis in a form which the engineer and administrator can use. The author has not had the opportunity to test the theory and data are not currenly available to apply it directly. It has not therefore been possible to provide case studies. But most of the necessary information is available to highway departments and could be assembled in a form appropriate for the application of the economic theory without undue difficulty. Such data as are not already available should not be difficult to collect. While arranged with problems of highway planning specifically in mind, the basic principles of analysis are equally applicable to economic problems which are faced in other public utilities.

In simplifying welfare theory sufficiently to make it comprehensible to the layman and practicable of application, some of the more abstruse theoretical controversies have been omitted from the text. A discussion of these will be found in the appendix. It is hoped that fellow economists will read this appendix before condemning the analytical techniques used. The non-economist who is prepared to take the economic theory on trust, however, can safely omit the appendix.

CHAPTER ONE

THE ECONOMIST'S APPROACH TO HIGHWAY PROBLEMS

The Peculiar Nature of Highways

The economist is concerned with the production of goods and services which have utility, or the power to satisfy human wants. Utility is a matter of having the right thing in the right place at the right time. Within the productive process we find activities concerned with creating the form of a good or service, transporting both the component parts and materials and the finished products to the place where they will be used, and storing them until needed. Transportation is vital to the whole productive process, and can be regarded as an industry. It is with transportation that we are concerned in this book, but highways themselves do not provide transport. They are an immovable good which provides services to the vehicles which do provide transport. So closely are highways connected with the things and people which move on them, however, that they must be treated as the fixed plant of the highway transport industry.

This apparently complex relationship of a good which provides a service to another which in turn provides the services desired is not an unusual one. We find it with railways and to some extent with airports and harbours. The peculiar problems which highway planning poses for the economist lie in the relationships between the highway and the vehicles and persons who use it.

Such relationships are normally of two types, and in a monetary economy two methods of apportioning costs have been developed to suit the different conditions. In the one case the good or service is delivered in known quantities to known persons, who pay a price for its use. In the railway industry every passenger or shipper of freight pays something towards the cost of the track facilitating the movement. Apportioning such costs equitably is a very complicated problem, but the persons who have derived benefit are known. Although more than usually complicated, the system of pricing whereby the user of the service pays for it is the same in principle in railroading as it is in any other industry. The nature of this financial transaction is obscured in the case of railways by the fact that the same organization normally controls both the track and the vehicles; but this does not change the essential nature of the transaction, which is the same in the case of harbours where the docks and the ships might be owned and controlled by different organizations. Our first case then is where the service is sold at a price in known quantities to known persons.

In the second case the service is provided to the economy at large and the identity of the users and the extent of their use of the service is not known. Normally such services are provided by government and paid for by taxation based not on the benefit derived but on some other basis. Thus our defence system is paid for by various taxes which do not attempt to measure for each individual how much he is defended. This type of relationship, involving governmental control, and financing through taxation, does not always hold when these methods of payment are used. In some cases the users can be identified and the extent of their use of the system is known, but the taxation system is used rather than the pricing system for reasons of equity or simplicity. In education it would be quite a simple matter to charge each pupil for the services received, as private institutions do, but many governments prefer to assess the cost against the community as a whole on grounds of equity. Similarly it would be possible to charge for garbage collection on the basis of amount collected, just as electricity is charged for by the amount provided, but it is simpler to assess the cost against users by a property tax. The distinction between the two types of relationship is not based on the type of financial system actually used—price or tax—but rather on whether the user and extent of use are determinable or whether the service is spread indiscriminately over the community as a whole.

The problems of highway planning arise from the fact that it does not fit into either of these categories. Everyone in the community uses highways in some way, but the degree of use varies greatly. This diversity of users is not peculiar to highways but applies equally to most public utilities. The difficulty arises because the services of highways are not delivered to users in the way that electricty, for example, is delivered. This makes it impossible to determine the amount of use made of the highway by each person, which in turn makes the use of a price system extremely difficult. At the same time highways cannot be grouped in our second type of relationship because their services are not spread evenly or indiscriminately over the whole community. Everyone uses roads and streets to some extent, but their use by motor vehicles, and particularly commercial vehicles, makes the use of a general taxation system of financing inappropriate. This is particularly important where road transport is in competition with other transport media. The economic balance between road and rail transport would be severely upset if highways were provided out of general taxation, but railways had to meet the cost of their track. The problem of financing highways and devising a system of assessing costs against the users is such that neither a simple price system nor a taxation system is appropriate. We shall examine it in detail in later chapters of this book.

The absence of a market in which users and the extent of their use are identifiable also poses severe problems in the planning of highways. The economic market serves two distinct but related functions. On the one hand, it allocates goods and services between consumers and recoups from them the costs of production; on the other hand, it guides production and allocates resources into the production of those goods and services for which consumers are prepared to pay most, which are assumed to be those from which they derive the greatest benefit. In the absence of such a market it is difficult to assess how highly the consumer values better highways, and there is therefore no simple guide to the allocation of an optimum volume of resources to the production of this good.

This absence of a market as a guide to the optimum allocation of resources is, of course, typical of those functions which fall within our second category. A municipal swimming pool might be planned as an economic proposition on the basis of anticipated numbers of users prepared to pay a certain admission fee, but where parks are provided out of public funds with no entry fee there is no such guide. Similarly, one cannot plan a defence policy on the basis of what people are prepared to pay individually to be defended. These problems are solved in a democratic community by politicians who are elected to make decisions for the community as a whole on the basis of what they consider the community is prepared to pay for certain services. This system works well where the good or service in question is of benefit to the community as a whole; and benefits are more or less evenly distributed. While this condition might be satisfied for local roads and streets it is not true of major highways where the benefits are predominantly for the users of motor vehicles. Road transport is an industry in competition with rail and other transport media, and a system in which railways are planned on market criteria and highways as community services severely upsets the balance of competition.

Because the entire road system is an integrated whole, it is not possible to separate major highways from local roads and streets in an attempt to plan the former on commercial principles and the latter on political estimates of community welfare. Every trip begins and ends on a local road but major highways might be used for most of the journey. There is, therefore, a joint demand for the use of major and minor roads, and the combination desired for each trip is different. Successful attempts have been made in the United States to isolate some major highways from the rest of the system and plan them on commercial principles, with a price charged for every trip in the form of a toll. Feeder roads are then regarded as a separate system linked with and providing access to and egress from the toll road in just the same way that roads serve railways. Such a system is of very limited applicability, however, and can be used only in cases where there is a large volume of through traffic which can be served adequately by widely spaced interchanges. When interchanges are necessarily frequent, or traffic volumes low, the cost of toll collection would become very high, and the many toll booths would constitute a serious impediment to the free flow of traffic. The problem therefore remains of most of the major highways which cannot be satisfactorily isolated from the rest of the road system.

While it is not possible to separate major highways from minor streets in an integrated road system for purposes of planning, it is possible to separate the control and planning of the highways from the ownership and operation of the vehicles which use them. This is normally done, and while vehicles are under private control the responsibility for highways rests with the various levels of government. This again emphasizes the unique relationship that exists between the trucking firm operating on commercial principles and the government agency which must provide it with highway services, although such an agency is best suited to the provision of services to the community at large.

Highways must normally be under the control of government for three reasons. In the absence of a market wherein users and the extent of their use of the highways are easily identified, it is impossible to use a pricing system appropriate to a commercial enterprise. This is possible in the case of toll roads, but the necessary conditions for isolation of one road are exceptional. In the second place, the acquisition of continuous right of way normally calls for compulsory purchase, which involves the exercise of authority not vested in any private concern. Finally, the economies of scale and indivisibility of highways make it impractical to have competing systems of roads, and the responsible authority must therefore be a monopoly. It is normally considered desirable that a monopoly which is responsible for services so vital to the existence of a community should be under public rather than private control.

The fact that some of the services offered by roads are vital to the existence of a community while others are predominantly in the interests of motor vehicle users has led to a great deal of confusion among writers on the subject. Attempts have been made to separate community uses from traffic uses in a classification of beneficiaries of the road system. The result is to create an insoluble problem of reconciling two different types of demand. Faced with this problem of a service which does not belong to either of the two types examined above, previous writers have attempted to identify the users belonging to each type.

The highway problem, however, is not a mixture of the two types but a third type which has some of the attributes of each, but which cannot be broken into its component parts. The only purpose of any road is to facilitate the movement of persons and things. We shall use the generic term "traffic" to encompass all movement on roads whether it be pedestrian or vehicular. Roads do not offer any services to the community which are distinct from the services offered by traffic. To attempt to separate services to the community from services to vehicles is therefore misleading when the former must necessarily follow from the latter. It is argued that gaps must exist between building for the access of light and air and as fire breaks, but such gaps are quite distinct from their use as a road. Back to back houses are no longer built for these reasons, but there is no public right of way across adjoining plots. Access to property calls for right of way across land not part of that property, but to maintain that this is a community use distinct from the passage of whoever or whatever requires access is a confusing misconception. Pedestrians are traffic, so are postmen, fire engines, ambulances, army convoys, and all the other forms of traffic normally called community uses.

Some ancillary services such as street lighting might yield benefit to persons not using the roads, but these are quite incidental just as noise from a railway is incidental. The sole purpose of street lighting is to serve persons using the street. Below the surface are found sewers, gas and water mains, and so on, and these can be regarded in two ways. Firstly, they can be regarded as separate from the highway, just as underground railways are separate, and their location purely one of convenience not one of necessity; or secondly, if they are held to use the highway in any sense it is as what they are, forms of traffic. It is true that some of these forms of traffic are so basically essential to the existence of a community that some form of right of way has been necessary since prehistoric times. But this only emphasizes the importance of land transport; it does not make these forms of traffic anything but traffic. Throughout the analysis which follows we shall therefore regard the sole purpose of highways as the provision of services to traffic, though there are of course numerous types of traffic that are by no means homogeneous.

The Nature of Highway Services

One of the fundamental principles of economic analysis is that a problem should be analysed within the context of the appropriate conditions. The analysis of a commercial enterprise is not appropriate to the analysis of a community service, and neither is appropriate for the analysis of highway planning. Attempts have been made to apply the principles of each to this problem, and we must therefore examine why they are not appropriate before proceeding to devise a suitable approach.

The community service approach was adopted in the days before the advent of the long-distance stage-coach. At that time roads served local pedestrian and animal traffic, and long-distance movement was effectively confined to military operations and other governmental movements. The benefits from roads were spread evenly through the community, right of way had been public domain since manorial times or before, and no expensive services were called for. Maintenance was provided for by statute labour which, however inefficient, did spread the burden equitably among the community. There was no impact on the balance of competition in the transport industry since there was little long-distance road transport, railways were unknown, and waterways served a different market. These conditions changed radically with the advent of the stage-coach and later the motor vehicle. Today the bulk of expenditures are necessitated for the service of motor vehicles, the benefits from which are not spread evenly throughout the community, and which do compete with other forms of transport. The attitude to highways as a government service to the community at large has persisted, however, long after the conditions which made it appropriate have ceased to apply. This is particularly true in England where attempts to modify the system by special motor vehicle taxes tied to a road fund failed, and the old idea was resorted to of a community service financed by the exchequer from general taxation with no connection between vehicle taxation and road expenditures. This is clearly not appropriate to the full role which highways play today.

The need for improvements in roads between settlements as distinct from those within towns came with the growth of stage-coach traffic. These roads would pass through sparsely populated country and serve no community uses. It was therefore felt that they were part of the commercial enterprise of road transport and a normal business approach was adopted, the roads being improved by private funds by turnpike companies which then had the right to charge tolls. This was a realistic approach while the traffic was easily identifiable and could be charged according to use. A similar approach is used today on some limited-access motorways in the U.S.A. Though it is questionable whether this is the best system of finance under these conditions, it is an appropriate one, but only so long as the traffic is identifiable and easily charged. As we have seen above, however, these conditions do not hold on the majority of highway or on city streets, and the great expense involved in attempting to collect tolls on all roads, where each trip has a different origin and destination, makes such an approach inapplicable under current conditions.

Thus, while the community service approach and the commercial approach are tenable under certain limited conditions, neither is adequate for the complex problems of modern highways. Instead, we must devise a third approach which treats highways as a public utility. Responsibility must rest in a government agency because a normal commercial system cannot operate in the absence of a market in which the service can be sold. Furthermore, highways offer services to the whole community of such a vital nature that it is expedient to vest their control in public hands. But at the same time they offer services to the commercial enterprise of road transport, in competition with other transport media, and this makes it essential that the criteria of planning and financing be analogous with those used by a commercial enterprise. The whole nexus of forces governing the planning and financing of a commercial enterprise is co-ordinated by a market, but there is no market for highway services. We must, therefore, examine what these forces are and how they are automatically co-ordinated in the free market, in an attempt to devise a technique of planning which will be based on the same criteria, but co-ordinated rationally by the responsible authority rather than automatically by a free market.

The Nature of Economic Criteria

There are three basic forces that govern economic activity in a capitalist society; the availability of resources, the tastes and preferences of consumers, and the command of consumers over resources. In non-capitalist societies attempts are made to modify the last two forces, but the first is fundamental to any economy. Tastes of consumers can be overridden by opinions as to the needs of consumers; and their command over resources, which depends on income distribution, can be controlled by taxation and subsidy or wage control. These modifications do not affect the interaction of the three forces, although their automatic reconciliation in a free market might be superseded by administrative planning. In all capitalist societies today a certain amount of control is excercised over the operation of free markets because it is considered that the free market does not always operate to the best advantage of the community as a whole. In the case of highway planning, however, administrative planning is necessary because a free market cannot work at all. Our problem is to devise a technique of planning, based on assessments of the basic forces, which achieves the same results as a free market would achieve if it could exist, but modified where improvements are possible, in the same way that railways are controlled in an effort to improve upon what would be the free market situation.

This task is fortunately not as impossible as at first sight appears. The development of micro-economic analysis has enabled us to examine the free market and find out, not only how it works, but also what resultant volumes of production and price levels will be achieved in equilibrium from any given set of forces. The hypothetical cases of pure competition and pure monopoly are relatively simple, and any reader not familiar with the theory of the firm under these conditions is recommended to refer to any elementary economics textbook for enlightenment if the analytical chapters of this book prove difficult to follow. Analysis of the more common market types of oligopoly and monopolistic competition becomes more complex and often results in indeterminacy. The transportation industry is one of differentiated oligopoly; there are few producers of services, and these services are not homogeneous. A free market of this type is characterized by indeterminacy. This, however, does not make our task impossible, because in practice a free market is not permitted in the transportation industry.

Analysis of the various market structures has enabled us to determine the most desirable features of each. The philosophy of the private enterprise system is that the most desirable results are achieved under conditions of pure and perfect competition, but these conditions are rarely found in practice. Monopoly and oligopoly permit various practices of price discrimination and profit maximization, which are controlled by various techniques of combines legislation. It is impossible to secure perfect competition because the economies of scale and the indivisibility of plant in many industries make it impossible to have many independent and competing firms.

In devising a technique of highway planning based upon the free market, we are essentially synthesizing a market. The market operates as an automatic computer to establish an equilibrium between the various forces. But each market type will achieve different results because the relationships between the various forces are different. In synthesizing a highway market we can combine the best features of the various market types to achieve the type of optimum solution for which legislators strive when they modify free markets by combines legislation, rate control on railways, and various other policies. Since it is the normal practice in contemporary society to employ such modifications, we, in devising a synthetic highway market, are justified in having all the modifications built in.

The three basic forces—availability of resources, tastes and preferences of consumers, and wealth distribution—will govern our planning technique in just the same way that they govern any free market. The tastes of consumers determine which products they value most highly, while wealth distribution allocates the purchasing power which enables the consumers to convert their desires into effective demand. Demand is the willingness and ability of consumers to purchase certain quantities of goods and services at given sets of prices. The producer faced with certain knowledge and anticipation of the demand for his product in turn has a demand for resources. The interaction of the demand for resources and their availability establishes resources prices, which in turn determine costs of production for the entrepreneur and income for the resource seller. The former determines the willingness of the entrepreneur to produce and sell certain quantities of his product at certain prices, while the latter determines the ability of the consumer to express his tastes and preferences in terms of effective demand. The interaction of these two forces, supply and demand, determines the prices and quantities of the various products produced and sold. In a free market economy the relationship between the basic forces and the various resource and product markets is extremely complicated. There is no temporal sequence in the above description; each market situation is simultaneously dependent on the others.

In devising a substitute for a highway market we do not have to construct a complete set of markets, but only one more to fit into the existing nexus. This will, of course, have impacts on other markets and repercussions back on the highway market itself, but the most important effects will be foreseeable and the minor effects negligible in importance.

Highways consume scarce resources, just like any other productive activity, but we already have an established set of resource markets with known market prices. A considerable change in the demand for the resources needed in highway construction would have an impact on these prices and therefore on other industries using them, but other markets would automatically readjust to equilibrium and the effect of changing prices on the highway construction industry itself could largely be foreseen. As far as the cost of construction in the highway industry is concerned, therefore, we have a situation comparable with any other industry. This does not take the form of a supply curve simply because there is no market in which to offer the product for sale. But since we know by analysis how a supply curve is derived in a free market from the cost data we can calculate what the highway supply curve would be if there were a market.

The demand for highway services is similarly governed in the highway industry by the same forces as in any other market. Consumers have tastes and preferences for good highways in relation to other goods and services, and purchasing power to enable them to express their desires in terms of their willingness to purchase certain quantities of highway service at certain prices. Again, this does not take the form of a demand curve because there is no market in which to express a willingness to buy. Major highway improvement programmes would have an impact on tastes and preferences by changing the availability of transport service and therefore the availability of other goods and services to the consumer; and an effect on wealth distribution resulting from changes in earnings following the impact on the labour market of new employment opportunities. But again, other markets would automatically come back towards a new equilibrium position and the impact on the highway market itself could be foreseen and estimated. Just as we know by analysis how a supply curve is derived from costs of production, so we know how a demand curve is derived from the tastes and purchasing power of consumers. Given the data we could therefore calculate what the demand curve for highway services would be if there were a highway market.

The gap caused by the absence of a highway market is not nearly as extensive as might at first have been thought. We have all the necessary forces which would determine both the cost and the demand curves, and by calculation we could discover from the relationships between them the appropriate volume of production and price. This acts as a guide to the planning authority as to the desirable extent of highway improvement programmes, while various forms of taxes and charges levied on highway users can be devised to collect the appropriate price.

While the idea of thus synthesizing a market is simple, in practice and in detail many problems have to be overcome. The simple supply and demand curves, and the significance of the location of their intersection, belong to a perfectly competitive market. Highways can never be that, and we shall instead have to use analogous techniques appropriate to conditions of monopoly or differentiated oligopoly. Since, however, we are looking for an optimum solution and our constructed market is synthetic, we can combine the analyses of different market structures so as to achieve results compatible with the necessary conditions for welfare maximization.

A more serious difficulty is that the unit of production is not homogeneous from either the production or the consumption standpoint. Highways cannot be planned in terms of so many lane-miles without reference to location. We must therefore face the problem not only of *how many* highways but of *which* highways. In addition, of course, highways can be constructed of various types and qualities. Similarly the demand for highway services does not exist in homogeneous units. Quite apart from the difference in demand for highways of various types in various locations, the demand for the service of a given highway to support one ton for one mile differs with each type of vehicle. A ton-mile by a heavy truck at 50 miles per hour is a different service from that for a passenger car at 65 miles per hour. Our theoretical substitute for a market must take account of these difficulties, and while the basic idea of a synthetic market is simple its application to such a complicated case leads to some complexity.

When we have devised the theoretical framework its application to a given case calls for precise data. The cost data are not difficult to assemble, but their conversion to monetary units can be a formidable problem in cases such as the cost of life lost in road accidents. Similarly the collection of data on consumers' desires and willingness to pay for highway services presents difficult problems of evaluation.

Finally, a system of taxes and charges which will serve the function performed by prices in the normal market must be devised so as to be both simple to administer and equitable in its incidence.

These three problems—the theoretical framework, the collection and evaluation of data, and the financing system—are examined in turn in the chapters which follow.

The Limitations of Economic Analysis

The object of this book is to provide a framework for decision-making. Many of the decisions which must be made are matters of opinion and the economist, as an economist, cannot offer advice on these. He must limit himself to a rational course of action based on given values. Many articles have been written attempting to justify better highways in terms of economic advantages from the standpoint of community welfare, defence, safety, an so on. In doing so the writers have made many value judgements which have no place in economic theory. Such arbitrary opinions have been attacked by welfare economists as being without rational foundation. Welfare economics has performed valuable service by destroying the invalid, but has substituted no alternative rational basis of decision-making which does not involve value judgments.

Both arbitrariness and rationality have their part to play in the process of decision-making, but the economist can help only with the latter. Arbitrariness implies determination at random or as a matter of opinon or discretion. As such it can be objectively neither right nor wrong. Rationality on the other hand implies a reasoned deduction from given data and defined ends and as such a conclusion can be held to be right or wrong objectively, according to whether or not it follows logically from the given premisses. For our purposes the availability of resources and prices in resource markets are given premisses, as are the tastes and preferences of consumers and the existing wealth distribution. These must be assessed, but they cannot be criticized by rational analysis. We can, however, devise an optimum solution in terms of welfare maximization from such given preferences by a purely rational process using the criteria of the free market, modified by current political practices, as a basis. Problems of data collection are largely statistical, the aim being to discover what the data are, not to offer opinions as to what they should be. Problems of valuation of data, as for example the monetary evaluation of death or injury in highway accidents, are largely matters of opinion on which the economist can offer no advice beyond the valuation of lost production. The human sacrifice is a matter to be assessed by law courts or politicians who are vested with authority to make such decisions. The application of our theory to specific cases will call for the evaluation of such data, but the administrator must look to persons other than the economist for guidance on these problems. The economist can offer advice on highway policy only within the framework of given data. Where these data can be collected techniques of statistics are involved. Where new decisions on values must be made for the community as a whole, the task rests squarely with the politicians.

As the British Minister of Transport remarked when speaking on the Commons Committee stage of the Road Traffic Bill on December 15, 1955, "Sooner or later . . . it becomes incumbent for better or worse on the Government to take a decision one way or another."

CHAPTER TWO

THE COSTS OF HIGHWAY TRANSPORTATION

THIS CHAPTER and the two which follow develop the theoretical framework of highway planning. In this chapter we examine the nature of costs, chapter III deals with the demand for highway transportation, and chapter IV with their reconciliation to determine the optimum plan. In using costs and demand as the two basic forces we are following the normal procedure of economic analysis. This differs, however, from the "cost-benefit" analysis which has been advocated by other writers on the problems of highway planning. The major difference lies in the product under consideration. Cost-benefit analysis treats the highway itself as the product and assesses and compares the cost of highway construction and the benefits derived in the way of reduced vehicle operating costs, time saved, etc.[1] This approach is a simple guide where a single modification to an existing highway system is under consideration, but it has severe limitations when applied to a major project or when we attempt to analyse the system as a whole. In order to assess benefits some estimate of traffic volumes must be made, but the volume of traffic, or demand for use of the road, is itself a function of the cost of road use, both in terms of vehicle costs and payments for highway use. When comparing two alternative plans by cost-benefit analysis we can either assume that the volume of traffic will be the same in either case, or we can assume that it will differ, depending on which road is built. The former assumption ignores the impact which better roads and reduced costs will have on the volume of traffic; the latter poses further problems. We can assess the benefit to existing traffic from reduced costs following improvements, but where new traffic is generated we have no way of assessing how much benefit it derives. This is not simply the difference between the old and new cost levels, because this traffic would not flow at the old cost level. Furthermore, if we accept the fact that the volume of traffic will depend on cost levels we have no way of determining what will be the optimum volume of traffic, and therefore no guide to a pricing or taxation system. This problem is examined later in this book and a brief example will serve at this stage to show its importance. Under congested conditions a motorist will make a trip if the value of the trip to him is greater than the costs of vehicle operation, taxes, time, and inconvenience which he will encounter. However, his presence on the road will increase congestion and raise costs for all other motorists, and if these costs were taken into account as well as the costs which the motorist in question bears, the trip might well be not worth while. The marginal cost of the trip if all costs are considered is thus greater than its marginal utility, the trip is not worth while, and the volume of traffic is therefore too great. A pricing system devised to ensure that only the optimum volume of traffic flows could increase the aggregate benefit derived from the highway. We have no guide to such a pricing system, however, unless we know what the optimum volume of traffic would be and cost-benefit analysis gives us no guidance on this point.

The analysis developed in this book, based on costs and demand, is concerned with the final product, highway transportation. The highways are in the nature of producer goods, or capital; they do not have utility in themselves but are used in the production of goods or services which do have utility. The costs of the highway are thus only one of the costs of the final service of transportation. Other costs include the costs of vehicle operation and the time, inconvenience, and risk of accident borne by the users. The measurable benefits of the cost-benefit analysis are treated as changes in costs in our analysis, and the complicated process of reconsiling costs and benefits, so as to maximize benefits in relation to costs, is reduced to the comparatively simple process of minimizing total costs. In itself this difference between the two techniques of analysis is no more than a rearrangement of figures by arithmetic manipulation. Its importance lies in the fact that by reducing the two concepts of costs and benefits to the single concept of aggregate costs we are able to introduce a further concept, that of demand, without making the analysis unduly complicated.

Demand measures the utility derived from transportation by the willingness of consumers to pay for it. This is a force of vital importance in highway planning, but is overlooked by cost-benefit analysis. The reconsiliation of costs and demand enables us to determine the utility derived from alternative plans and therefore to select the plan which offers maximum utility in relation to cost, or in terms of welfare economics that plan which maximizes aggregate consumer surplus. Similarly, it makes possible the determination of optimum traffic volumes and the devising of an appropriate pricing system. The function of the pricing system is to change the incidence of costs. The costs fall naturally on three distinct groups; highway costs on the highway authority; vehicle costs and the costs of time, inconvenience, and risk on the user; and certain miscellaneous items on the community at large. The function of the pricing system is to recoup from those deriving utility from the highways those costs not automatically borne by them.

The meanings attached to the terms "benefit" in cost-benefit analysis and "utility" in this analysis are quite distinct. Writers on cost-benefit analysis have often aimed at the concept of utility in their definition of benefits, but the technique of measuring benefits invariably limits the concept to reductions in vehicle operating costs and some of the users' personal costs of time, inconvenience, and risk. Utility, however, measures the satisfaction derived from travel whether that satisfaction is in a finished form, as in the case of pleasure travel, or a means to a different form of satisfaction, as in the case of commercial movements. "Benefits" have thus sometimes been taken to mean a measure of cost reductions; but "utility" is quite distinct from costs, just as the flavour of an apple is distinct from its price. The importance of the concept of utility to highway planning can be seen where a new road is planned, which will generate completely new traffic. Costs of all types will be incurred if this highway is built, and cost-benefit analysis is useless as a guide to whether such a project is worth while. The only way to determine this is by consideration of whether the utility derived from use of the road is so great that users would be prepared to pay more than the costs involved to derive this utility. The road might yield great utility, but no "benefits" in the sense of cost reductions. This effect is present to some extent in any project that will encourage the growth of traffic volumes, and the inability of cost-benefit analysis to cope with it is a major limitation which justifies the development of the cost and demand analysis.

The Nature of Costs

In the field of highway planning, characterized as it is by the absence of a market, determination of costs is in some cases a difficult problem. Discussion of the techniques of evaluating the various costs is deferred until a later chapter, but it is important at this stage to examine the basic principles of cost determination. The cost of producing any economic good is the total cost of all the resources which enter into its production. Economic resources are necessarily scarce, and to use a unit of a resource for one purpose means that it is not available for another purpose. The true cost of using a unit of a resource is therefore the sacrifice of the utility which it could have provided in the use foregone. This is the concept of opportunity cost which is basic to economic analysis. While we do not have a market for highways or highway transportation, the resources which are concerned in providing these services are the same types of resources which could have been used in the production of goods and services for which there are markets. The cost of highway transport is therefore the total value of the goods which could have been produced instead with the same resources. Without knowing what these goods would have been we can determine the opportunity cost of the highway by examining the components of their value.

The value of any good is composed of three parts: the cost of resources used; the implicit cost of entrepreneurial effort or, as it is usually called, a normal rate of profit; and pure profits. In so far as resources used for a highway are purchased in a free market the first component of opportunity cost will be included in highway cost. No account need be taken of the normal rate of profit which could have been earned by an alternative use of these resources, since this is defined as the opportunity cost of the entrepreneurial effort which would have been employed in the alternative use. This opportunity cost is the amount which that effort could have earned in some other line of endeavour. When the resources in question are used for a highway the alternative user will presumably follow that other line of endeavour and still earn a normal rate of profit, which is therefore not part of the opportunity cost of the highway.

The third component, pure profits, could arise from an alternative use of the resources used for the highway in only one of two ways; their use in the production of a good to be marketed under conditions of perfect competition in disequilibrium; or their use under conditions of imperfect competition. In the first case, the pure profits would disappear anyway as the market moved towards equilibrium, so we need consider only the case of an alternative use involving some degree of monopoly. The foundation of such monopoly power could rest either on some unique characteristic of the resources used, the scarcity of which precludes extensive competition; or on some other basis such as patent rights. In the latter case the monopolist will continue to make abnormal profits by replacing the sources used for the highway with others of similar characteristics. In the former case, where the basis of pure profits rests in the unique nature of the resources used for the highway, the existence of pure profits is fictitious since they will be absorbed in the market value of the resources in question. The market value cannot be less than the amount for which the seller is prepared to sell. Where the foundation of a monopolist's high profits rests in the resources being sold he will not voluntarily sell them for an amount that is less than enough to compensate him for the loss of these profits. Thus, under these circumstances the return in excess of a normal rate of profit is in the nature of a return on resources of high value rather than a pure profit. If the resources are used for a highway, the market value will therefore take account of all that could have been earned from any alternative use of them.

Where the resources used for a highway are purchased in a free market the value of the resources therefore measures the opportunity cost of their alternative employment.[2] The only common exception to free market purchase of resources is the compulsory purchase of land. In this case opportunity cost is still fully reckoned as the cost of the highway, providing the cost of the land is regarded as not less than the amount for which the seller would be prepared to sell. The problem of assessing this in practice, in a situation where the seller attempts to secure a price higher than the minimum for which he would be prepared to sell, is discussed in a later chapter.

While the cost of the highway, reckoned as the market value of the resources used, cannot be less than its opportunity cost, it might well be greater. This could occur, for example, in the case of land of unusual characteristics previously marketed under conditions of duopoly and monopsony, that is, where there are two sites suitable for the monopolist's purpose but he is the only buyer. Here the rent he pays will be an amount slightly less than the minimum amount for which he could acquire the site whose owner has the higher minimum selling price. This rent will be a free market rent before the highway project is considered. However, if the highway project involves compulsory acquisition of the land he occupies there will then be only one site available, and the market will be one of bilateral monopoly. The owner of this other site will then be in a much stronger bargaining position and the rent will rise. Accordingly, the minimum amount which the present occupier of the land to be used for the highway would voluntarily accept to move will be, not its market value before the highway project, but what will be the market value of the alternative site after the acquisition of his present site for the highway. Reckoning this amount as the cost of the land for the highway exceeds the opportunity cost of the land used, the excess going to the owner of the second site, who can now let for a high rent land which was previously, and would but for the highway project have remained, idle. This is but one simple case of the ramifications which a highway project can have on other land values.

Such possibilities need not concern us unduly, however, for they are typical of changing market conditions. There is no essential difference between the above case and the fact that civil engineers generally are likely to be better off as a result of an extensive highway construction programme, because the additional demand for their services raises their salary level in all related fields. The cost of their services to the highway project might well exceed their opportunity cost, but this is only one example of the way in which any economic activity is likely to benefit numerous people. It does, however, point up one important fact which must be borne in mind. When assessing the cost of resources used for highway construction at market value, the appropriate level is not the market value of such resources before the project begins, but what the market value will be when the impact of the additional demand is felt in resource markets, for this is the amount that will have to be paid. While this figure might well exceed the opportunity cost of using resources for highways, especially if the project is an extensive one, it certainly will not be less.[3] Thus no separate item of opportunity cost need be included in our cost analysis beyond the market value of the resources used.

So far we have considered only the explicit costs of the highway itself. These are by no means the total cost of highway transportation, however. The economic analysis of costs distinguishes explicit from implicit costs, but regards both as components of total costs. In this the economist's approach differs from the accountant's, the difference being important in that many writers on cost-benefit analysis have followed a procedure analogous to the accountant's approach rather than the economist's. This is not a matter of controversy between accountants and economists. Each uses a method appropriate to his task, but the appropriate tool for highway planning is the economist's rather than the accountant's.

Explicit costs represent monetary outlays by a firm on the purchase of resources and are recognized by both the accountant and the economist. Implicit costs represent the use of resources owned by the entrepreneur for which no monetary outlay is necessary. Since the accountant is concerned with the financial operations of a business he need take no account of implicit costs. The result of the accountant's work however, by ignoring implicit costs, does not present the complete picture. This can be seen by a simple example. An entrepreneur operates a drug store with his own capital and manages it himself. On a year's operation his revenues exceed explicit costs and he makes a certain profit. If we include in the profit and loss statement the implicit costs of the interest his capital would have earned if loaned to an undertaking with similar risks and the salary which the proprietor could have earned with similar work managing a drug store which he did not own, his pure profit is considerably less than his net profit in the accounting sense, and might even be a pure loss. In the latter case his operation of his own drug store is not an economically sound proposition, although his accounts by ignoring implicit costs show him to make a profit.

An analogous position exists in the highway field where there are considerable implicit costs for the motorist. The costs of vehicle operation are normally included in cost-benefit analysis in so far as they change as the result of a highway improvement, but the more personal costs of time, inconvenience, and risk of accident are not adequately reckoned. If we are to determine whether a highway project is economically sound it is important to take cognizance of all costs, both explicit and implicit. These can be conveniently classified.

The Classification of Costs

The costs of highway transportation can be conveniently classified according to their immediate incidence into four groups: (1) the costs of the highway itself including construction, maintenance, and operation, which fall initially on the highway authority, together with those opportunity costs such as property taxation which would rest on the highway authority were they levied; (2) the costs of vehicle operation which fall on those responsible for the operation of the vehicles; (3) the users' personal costs of time, inconvenience, and risk, which fall on the persons travelling or the owners of goods travelling; and (4) the costs which fall on the community at large or sections of it. In order to separate traffic from the other motives which might support highway construction, such as unemployment relief, the former is treated separately as the demand pattern, while the latter are regarded as negative costs.

For purposes of analysis we can distinguish fixed costs from variable costs by the traditional short-run analysis of the costs of a firm. The product is highway transportation and the time period the life of the highway. Some of the costs will vary with the amount of transportation provided while others will not. These are our variable and fixed costs respectively. The variable costs cause considerable analytical difficulties because of the absence of a single measure of units of transportation, but at present we can regard any cost which varies in any way with the amount of transportation as a variable cost. The important distinction for purposes of planning analysis is between fixed and variable costs, while the distinctions of incidence are important to the formulation of a pricing policy. We must now examine the components of each category of costs. Full discussion of the problems of evaluating these costs is deferred until chapter v.

Fixed Highway Costs

Right of Way

The value of land cannot be separated from the value of buildings on it and the cost of right of way acquired for highway purposes therefore includes all structures on it. The opportunity cost of using the land and buildings for a highway is the sacrifice of alternative uses. The monetary value of these is the most that some other user would be prepared to pay for them, that is, their current market value. The capitalized cost of right of way is, therefore, the sum of the capitalized current market values of all the interests held in the land and buildings acquired. By using a capitalized value rather than annual rental value, problems of property taxation are kept distinct. This capital sum can then be amortized over the anticipated life of the highway. The resale value of land should the highway be abandoned will be considerably lower than its value at acquisition, because buildings or soil fertility will have been destroyed and the pattern of adjacent land ownership and use changed. If the potential scrap value is significant the initial value can be amortized down to scrap value over the anticipated life of the highway, the cost of continued use of the right of way after the anticipated life being the rental value of the land in its condition at the time or the interest on the resale value, whichever is higher. This arises from the fact that opportunity cost is the sacrifice of the best alternative use, in this case either letting the land or selling it and deriving interest on the capital. Anticipating the life of a highway is difficult. The initial intention is that it will continue in perpetuity and few highways have in fact been abandoned. However, traffic patterns and forms of transport change and in view of the difficulty of long-term forecasting it is normally considered prudent to anticipate a life of twenty-five years at most, or less under particular conditions as in the case of a mining road where the mineral resources will be exhausted in a few years.

Development Costs

This group of costs consists of those items necessary to convert right of way from its previous use to a condition suitable for the construction of a highway. They will be governed partly by the type of highway to be built, and partly by terrain, soil conditions, climate, and so on, but not by the extent of use of the highway once it is built. Though they might require future expenditure for maintenance they will never need replacing. The border line between replacement and maintenance is an arbitrary one, but is usually clear. It does not matter from the standpoint of this analysis on which side of the line doubtful cases are deemed to lie, but items that will need replacing are classified as construction costs rather than development costs. The major items of development costs are demolition, grading, and subgrade work (soil stabilization, etc.), together with intangibles such as legal costs. Each item will be composed of expenditure on management, labour, materials, equipment, fuel, insurance, and so on. The development costs can be regarded as a once-for-all capital expenditure to be amortized over the anticipated life of the highway.

Construction Costs

This category of costs is composed of the long-term capital expenditures on items that will eventually require replacement. It includes the cost of base and surface of the roadway, structures such as bridges and grade separations, drainage, and engineering. Each of these items can be amortized over its anticipated life or the anticipated life of the highway, whichever is shorter. An arbitrary distinction is made between development costs and construction costs where the necessity for eventual replacement is questionable. A bridge, for example, might need replacing while a tunnel will not. Where the need for replacement is doubtful, the anticipated life will be more than the anticipated life of the highway, and it then makes no difference whether the item in question is regarded as a development or a construction cost.

Many items of construction cost will deteriorate even in the absence of traffic, but they will deteriorate faster the greater the volume of traffic. In estimating anticipated life to arrive at the annual fixed cost of amortization, it is therefore necessary to distinguish the fixed costs from variable costs. The annual fixed cost is that amount which would amortize the item over its anticipated life in the absence of traffic. The variable cost is that additional amount required annually to amortize the item over the shorter anticipated life resulting from its use. Thus if a surface would last ten years without traffic, the fixed annual cost is that amount needed to amortize construction cost over ten years. If a certain level of traffic would wear out the surface in eight years the annual variable cost is that amount which, when added to the annual fixed cost, will amortize the construction cost in eight years. If after five years the volume of traffic increases so as to wear out the surface in the next two years the annual variable cost will increase to that amount which, when added to the annual fixed cost, will amortize the outstanding portion of construction costs in a further two years. Thus the annual amount of fixed cost remains constant while the variable cost is a function of the volume of traffic.

Maintenance

The distinction between construction and maintenance is a somewhat arbitrary one. There is no logical justification for regarding the resurfacing of a highway as a construction cost and the repainting of a bridge as maintenance. The distinction between the two types of cost is not of analytical importance, however, but simply one of practical convenience. The items of maintenance are numerous, comparatively small, and often difficult to identify in advance, and it is therefore convenient in practice to group them as a separate category of cost. The important distinction is between those maintenance costs which vary with the volume of traffic and those which do not. The principle for distinguishing fixed from variable maintenance costs is the same as that used for construction costs. Those expenditures on maintenance which would be necessary to keep the highway in good condition if no traffic used it are fixed costs, and additional expenditures necessitated by use of the highway are variable costs. Some items, such as maintenance of fences, signs, hedges, ditches, and snow clearance, will not vary at all with the volume of traffic, while surface maintenance will. Annual maintenance expenditures will increase with the life of the highway and to arrive at an annual cost figure it is necessary to discount the anticipated future stream of maintenance costs to a present capital sum and then amortize that sum over the anticipated life of the highway. The annual fixed maintenance cost will then be constant, while the annual variable maintenance cost might vary over time with changes in the volume of traffic. It is possible to estimate future maintenance costs for any volume of traffic and by discounting and amortization we arrive at an annual figure. With such a figure calculated for a series of traffic volumes the annual variable maintenance cost is easily determined when the volume of traffic is known. As with development and construction costs the annual maintenance cost is quite distinct from the actual expenditure in any one year. The cost figure is a discounted average of an irregular stream of actual expenditures.

Administration

The administrative costs of the highway authority are governed by the extent and type of the highway network, but are almost entirely independent of the volume of traffic which actually uses a given network. They are therefore fixed highway costs. The fact that the network built is itself largely dependent on traffic volume does not affect the fact that the administrative costs for a given network are independent of the volume of traffic.

Property Taxation

It is often argued that part of the opportunity cost of a highway is the loss of property tax which would have been paid by any other user of the land. This argument is invalid for the community as a whole for the same reason that it is unnecessary to count the normal rate of profit of any alternative user as part of the cost of the highway. The alternative user will now occupy another site, make his normal profit there, and pay his property tax there. Furthermore, if we accept the above argument it is necessary to include not only the loss of property tax on the land actually used for the highway, but all changes in property tax yield which result from changes in land and property values caused by the construction of the highway. It is more convenient for our purpose to examine these changes in other land and property values separately, and they are included in community costs.

While we reject the above reason for counting property tax as part of the cost of the highway, we must at the same time reject the normal reason for its non-payment, namely, that it would only involve the community paying itself a tax and can therefore be ignored. Once we reject the community service attitude towards highways this argument fails because the paying and receiving communities of any property tax on highways are different.

Property taxation must be included as part of the cost of a highway in our planning analysis, not for the reason rejected above, but because we are trying to determine how much of available resources should be devoted to highways. Highways and other uses are essentially competing for resources, which should go to that use where the utility yielded is greatest. Industry measures this yield in terms of profit, and the profitability of alternative uses determines the amounts which users are prepared to pay for resources, or their market values. If highways are to compete for these resources on an equal footing, determining net yield by our cost and demand analysis, it is important that highways be subjected to the same overhead costs in the form of taxation that apply to other resource uses. Certainly the present system in the United States, whereby railways pay property tax but highways do not, is analytically indefensible and tends to upset the balance of competition between competing transport media. Property taxation at prevailing rates must therefore be included as a cost of the highway in our planning analysis. Whether it is actually paid or not is a separate problem, but its non-payment involves a community subsidy to highways.

VARIABLE HIGHWAY COSTS

This category of costs includes all those expenditures of the highway authority that vary with the volume of traffic. Depreciation of the highway (construction costs) and variable maintenance costs have been discussed above in the separation of these items from fixed costs. The remaining variable highway costs are those directly concerned with traffic, such as control systems, traffic lights, and police.

The difficulties in the analysis of variable highway costs come not in classifying them but in determining with what measure of traffic they vary. Some are a function of aggregate traffic volume while others depend on traffic density. This problem is discussed at the end of this chapter after the examination of other variable costs.

VEHICLE COSTS

The costs of vehicle operation are normally divided into fixed and variable costs depending on whether they vary with mileage. This distinction is important for the vehicle operator but not for purposes of highway planning. We are not concerned with the mileage travelled by any particular vehicle, but with volumes and densities of traffic. These can be achieved with various combinations of numbers of vehicles and average mileage per vehicle. The distinction between fixed and variable costs is dependent on the time period. In the long run all costs are variable. The time period of this analysis is the life of the highway which is considerably longer than the life of a vehicle. For our purposes therefore, all vehicle costs are variable. Improvements in highways can bring changes in all vehicle costs. Faster average speeds save fuel and time, and by permitting the same amount of work to be done with fewer vehicles save on depreciation, interest, insurance, and the like. Traffic density also affects average speeds and vehicle operating costs and, for a given highway, vehicle costs are therefore dependent on traffic volume and density. Traffic volume affects the number of vehicle miles travelled and therefore the over-all vehicle cost, while traffic density or congestion affects the cost per mile for a given vehicle.

Vehicle costs are self-explanatory and need only be listed. They include: capital cost of vehicles, amortized over their life and thereby including interest and depreciation; maintenance; garage; insurance; fuel; lubricants; tires; and brakes. Drivers' wages are analysed separately under users' personal costs. In assessing these costs net or gross of taxation care must be taken to distinguish those taxes which are sumptuary in nature and those which are charges for highway services. The former are taxes which must be borne by any industry, such as excise tax and sales tax, and must be included for the same reasons that property tax on land was included. The latter, such as fuel tax, are not costs of highway transportation but charges designed to shift the incidence of costs borne initially by the highway authority, and these should not be included in vehicle costs. The distinction between charges for highway service and sumptuary taxes is analysed in the latter part of this book (p. 147).

Users' Personal Costs

In addition to the highway and vehicle costs, road transport involves certain costs borne directly by the traveller or highway user. These must clearly vary with the volume and density of traffic since they arise only with use of the highway. They are the costs of time, inconvenience, and risk. Time includes that of all travellers, drivers, passengers, and pedestrians, whether employed at the time or not, together with the time cost of goods transported by road. Inconvenience covers the fatigue, irritation, and discomfort of travel—factors which vary considerably with the type of highway and degree of congestion. Similarly the risk of accidents to persons and goods is a direct function of road and traffic conditions. The extremely difficult task of evaluating these costs is discussed in chapter v.

Community Costs

This is the residual group of effects, beneficial and otherwise, that a highway brings to people other than those using it or responsible for vehicle or highways costs. It includes some items which vary with the volume and density of traffic and some which do not.

The fixed community costs include loss of amenity which a new highway may bring to a beauty spot and the barrier which a limited access highway may cause. Some of these costs will be negative, such as its availability for use in wartime, which has a continuing value irrespective of its actual use for defence. Other costs will vary with the extent of use of the road: for example, changes in costs on other roads will be a function of the type and use of the road in question. Betterment and worsement of land values brought about by greater accessibility or noise and smell nuisance will also vary with the volume and density of traffic. The extent of these costs, both positive and negative, and how far they should be included in the planning analysis gives rise to many controversial issues. Discussion of these is deferred until chapter v which is concerned with the evaluation of costs. Some of the community costs must be included, however. and provision for them is therefore made in the following analysis.

The Analysis of Costs

For purposes of analysis the above costs must be expressed in precise terms as functions of the variable volume and density of traffic. This is a technique analogous to that used for the construction of the cost curves of a firm which express costs as a function of the volume of output. The individual highway or section of highway under analysis corresponds to the productive plant of the firm, but the analysis of the costs of transportation gives rise to problems peculiar to this case.

All costs must be reduced to a common unit of money. For some costs such as the highway and vehicle costs, this is comparatively easy; for others like the users' personal and community costs, it is extremely difficult. To avoid interruption of the analysis at this stage, discussion of the technique of evaluating the various costs in monetary terms has been deferred until chapter v.

Measurement of the volume of output involves two problems. The first is that vehicles are not homogeneous. The costs of a trip over the highway vary with the type of vehicle, whether it be a light or heavy truck, passenger car, or bus. It is theoretically possible to proceed with the analysis using a classification of vehicles and the numbers in each class, but it is the intention of this book to express the analysis of the fundamental issues, by use of comparatively simple geometry, and to avoid the other more complicated mathematical techniques that such detail would involve. If such detail is needed for practical application of this analysis, it can be handled by restating the theory in more complex mathematical form.

We overcome the problem here by using a "standard vehicle unit," the various types of vehicle being reduced to an equivalent number of standard units. The standard unit will have the same effect on all costs whether it be a unit of passenger cars, trucks, or buses. Devising a formula for reducing the various types of vehicle to standard units involves considerable difficulty as the relative impacts of different vehicles on the various costs differ. Thus a slow truck might cause lower highway costs in wear and tear than a fast bus, but bring about greater users' personal costs by causing congestion. These problems can be overcome only by a compromise formula, and we shall assume that such a formula is possible. This assumption does not affect the validity of the theory; it is made only to simplify the exposition of the analysis.

The second problem involved in measuring output is that transportation is not storable. If we could move vehicles in an evenly spaced stream day and night the total costs of a given number of vehicle units per year would be considerably lower than they are with the irregular stream that occurs in practice. Costs therefore vary not only with the total size of the traffic stream but with irregularities in the rate of flow. To overcome this problem we must express the costs of highway transport as a function of two variables, volume and density. Volume of traffic refers to the total amount of traffic in standard vehicle units passing a point in a finite period of time, normally a year. Density of traffic refers to the rate of flow of traffic past a point at a moment of time, and is measured in standard vehicle units per hour. The more marked the seasonal or daily peaks in traffic, the more irregular will be the traffic density for any given volume. While some costs of pavement wear are functions of the aggregate volume of traffic, others, notably those costs associated with congestion, are entirely a function of density.

Volume and density of traffic can be measured either separately for each direction of movement or as the sum of movements in both directions. The appropriate technique will depend on conditions of the project under analysis. On a divided highway it is more appropriate to consider the traffic flows independently, because costs on each carriageway are almost entirely independent of traffic volume and density in the opposite direction, and because solutions might emerge from the analysis which could not be analysed in terms of a combined traffic flow. Thus, where the daily peak traffic in one direction is spread over a shorter period than in the opposite direction, thus resulting in daily volumes in each direction that are approximately equal but with different peak densities, it might prove desirable to have more lanes in the direction of the highest density peak. The author does not know of a case of a divided highway where the numbers of lanes in each direction are unequal, but on some multi-lane urban expressways the central lanes are reversible in direction to achieve this in peak hours. Such a possibility can be analysed only by examining flows in each direction separately. On a two-lane highway, however, vehicle and users' personal costs are dependent on traffic density in the opposite direction as well as in the direction of travel because of the effect on overtaking. Here it might be more appropriate to proceed with the analysis in terms of the combined volume and density in both directions.

A further problem then arises that cost levels depend not only on aggregate density but on the directional distribution of that density. This is particularly true on three-lane highways. To allow for this would involve considering densities in both directions simultaneously but independently. This would pose serious technical problems in the simple geometric analysis used here, but as with other refinements would present no insurmountable difficulties if other mathematical techniques are used. Accordingly, we shall proceed on the assumption that volumes and densities are reckoned either in one direction or in both directions combined and defer simultaneous but independent reckoning to a mathematical restatement of the principles of analysis. Some costs are predominantly a function of volume while others are a function of density. We must now examine the costs in turn to see how they fit into this analytical framework.

The fixed costs cause no difficulty since they are independent of traffic volume and density. The sum of fixed highway costs and fixed community costs for the road under analysis gives total fixed costs which can be shown either as a horizontal line, or more usefully as an average fixed cost curve where average cost is total annual fixed cost divided by the annual traffic volume. Such a curve is of course a rectangular hyperbola. (See diagram 1a).

The variable highway costs are a function of both the volume and density of traffic. Depreciation and maintenance are largely determined by traffic volume, but they might be affected to some extent by density, in that frequency of impact load as well as number of impacts determines surface deterioration. Other variable highway costs such as traffic control systems are entirely a function of traffic density. The costs that are a function of volume are plotted as an average

Diagram 1*a* Diagram 1*b*

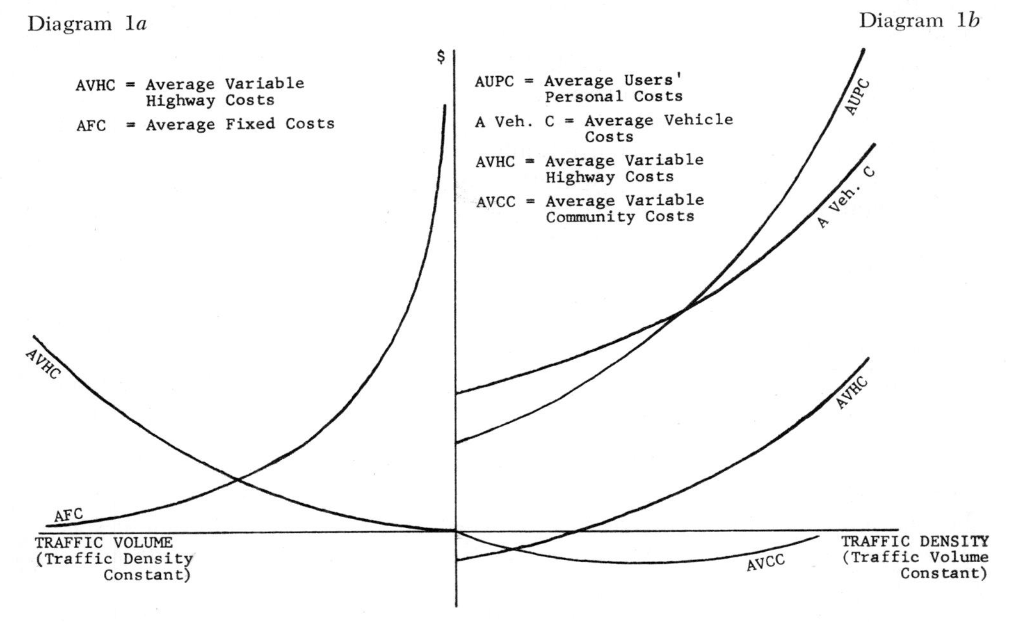

variable highway cost curve on diagram 1*a*. They are necessarily zero with no traffic and rise slowly at first, as a highway will accommodate a certain amount of traffic before it begins to deteriorate significantly faster than it would with no traffic. As traffic volume increases variable highway costs rise at an increasing rate and the average will therefore rise. Costs on diagram 1*a* are plotted on the assumption that traffic density is a constant. This unrealistic assumption enables us to isolate those costs which are entirely a function of volume. Irregularities in density for a given volume increase the costs of depreciation and maintenance, the discrepancy from the norm of constant density being plotted on diagram 1*b*.

Diagram 1*b* shows the costs of traffic flowing at any one time as a function of traffic density with an assumed annual traffic volume. The variable highway costs of depreciation and maintenance increase as density increases. At that density which is the average density for the assumed annual volume the average cost of depreciation inflicted by each unit of traffic will be that shown in diagram 1*a*. At greater densities the average cost inflicted by each unit will increase, and at lower densities it will decrease. These discrepancies from the norm are shown on diagram 1*b*, being zero for that density which is average for the assumed traffic volume, positive for higher densities, and negative for lower densities.

The remaining variable highway costs, those associated with traffic control, are entirely a function of traffic density. These variable costs of traffic control are distinct from the fixed costs of traffic control devices. Thus laning, traffic dividers, traffic lights, and so on, are part of the fixed costs of the highway since once installed their costs do not vary with traffic density. The highway or projected highway under analysis will have the extent of such devices stipulated, and they are part of the fixed highway costs. Within any one general plan numerous detailed modifications are possible in items such as these, and the technique of deciding which of such devices in worth while is discussed in chapter IV. At present we are concerned with the analysis of a complete highway or project in which all such matters are stipulated. The variable costs of traffic control are those which are incurred only at certain times of high-traffic density, such as by traffic police. At low-traffic volumes only the minimum amount of such operations will be necessary, and since these are needed at all times they are a fixed cost. As volume grows, more police are needed and these extra requirements are therefore variable costs. The average cost per vehicle will therefore rise with traffic density.

Average vehicle costs on a given highway under given climatic conditions are a function of traffic density. The average vehicle cost is the cost of operating one standard vehicle unit for one trip over the highway. This will be a minimum at very low traffic density. As congestion increases vehicle operating costs rise at an increasing rate.

The average users' personal costs of time, inconvenience, and risk will similarly increase with traffic density, but will rise faster than average vehicle costs as they are more affected by congestion. It will cost twice as much in time if speeds are halved, and maybe more still in frustration, but vehicle costs, though increased, will be less than doubled.

The variable community costs are of three types. The impact of highway improvements on land and property values is considerable, but as we shall see in the full discussion of this point in chapter V these changes in land values arise only from a shifting of the incidence of vehicle benefits. While betterment and worsement of land values might be important as a taxation base, the planning analysis with which we are here concerned takes account of the immediate changes in benefits to traffic through the cost and demand pattern. Since it is these same benefits which are shifted to cause changes in land values, we must count them only once. We do not therefore include changes in land values as a community cost as this would involve counting them twice. The noise and smell nuisance of traffic to adjoining property is a community cost and this will increase with the density of traffic. It is one component of the variable community cost curve in diagram 1*b*.

The third type of community cost is the change in vehicle and users' personal costs which occurs on other highways as a result of traffic diversion to the highway improvement under analysis. The extent of this change is largely dependent on the density of traffic on the improved or new road, but it also depends on the source of that traffic. The traffic that flowed on the highway before improvement will presumably use it after the improvement, and this will have no effect on other roads. If the road in question is completely new there will be no traffic of this type. Traffic that is generated entirely by the new road and would not have flowed at all but for the improvement will similarly not affect costs on other roads. Traffic diverted to the new road will, of course, reduce cost levels on the roads from which it is diverted, and the extent of this reduction is a negative community cost of the road in question. Secular growth in traffic over time will contribute to traffic density. The extent to which this causes a change in costs on other roads is limited to that traffic increase on the new road which would have occurred on other roads but for the improvement under analysis. These various sources of traffic are further discussed in the next chapter as the components of demand.

Traffic faced with the alternatives of using the improved road or another road will base its choice on costs, taking into account those costs, vehicle and users' personal costs, which the traffic bears. If average cost on the new road is lower traffic will divert to it; if it is higher traffic will divert from it. The equilibrium position will be where the cost levels are equal. Thus our measure of costs as a function of density in diagram 1*b* is also a measure of cost levels at the same time on alternative roads. The average saving to traffic on other roads will then be the difference between this cost and what the cost would have been on the other roads but for the improvement. Assuming the latter to be known the saving becomes a function of the average cost on the new road which is in turn a function of traffic density. We can therefore express these negative community costs on diagram 1*b* as a function of traffic density.

At zero traffic density there can be no diverted traffic, so the saving is nil. At moderate density there will be some diverted traffic and some savings on other roads. The curve will therefore fall from zero and become negative for moderate traffic densities. As density increases further, cost levels on the new road will rise. The probable extent of generated traffic and secular growth then becomes greater, thereby increasing average cost on the new road, and therefore on the old road since the cost levels tend to be equal. The higher cost levels on other roads then reduce the savings from what they were at lower densities on the improvement in question. Further, since we are averaging these savings over traffic on the new road, the greater this is, the lower the average saving. The curve will therefore rise towards the zero axis as negative average community cost decreases. It cannot rise above the zero axis, however, unless cost levels on other roads are higher than they would have been but for the improvement. This cannot be due to secular growth which would have taken place anyway; and it cannot be due to diversion from the improved road unless cost levels are higher than they would have been without the improvement. In this case the "improvement" would have been a deterioration. But the effect of generated traffic on the new road could be to increase cost levels on other roads above what they would have been but for the improvement.

This effect could arise where the generated traffic follows a route which includes other roads as well as the improved road. The effect is well known where an improvement on one section of road causes a bottleneck and congestion at each end. The additional costs of this congestion are community costs of the improvement and might outweigh the other negative community costs mentioned above. Further projects to relieve the congestion would then count the relief as a reduction in vehicle and users' personal costs. Where projects involving parts of a route in this way are under analysis it is important to combine the results before making decisions, as the losses on one section might be more than offset by gains on others. Plans for inter-city motorways in England were delayed for some years because of the belief that they would only make matters worse by increasing congestion in the cities at each end. Though doubtful, this might possibly have been true while the motorway was considered in isolation, but had the project been considered together with projects for urban expressways or bypasses at each end the decision would probably have been otherwise.

The general shape of the average variable community cost curve is shown in diagram 1*b* and includes both these costs, and the costs of noise and smell nuisance which have been added to avoid the necessity for a separate curve.

Combination of the Cost Curves

The vertical addition of the curves shown in diagram 1*b* gives the average variable cost as a function of traffic density. As shown in diagram 1*a* there are also average variable costs as a function of traffic volume. These two parts of average total variable costs cannot be combined into one two-dimensional curve as there are three variables involved. They can, however, be combined into one three-dimensional surface. Diagram 1*b* is drawn on the assumption of a given volume of traffic and the average variable cost appropriate to this volume can be read from diagram 1*a*. If this is added as a constant to the sum of the curves in diagram 1*b* we have a cross-section of the three-dimensional surface at that assumed volume. By basing diagram 1*b* on a series of assumed volumes we could derive a series of cross-sections which combined would give the average variable cost surface. The significance of this surface begins at a volume of one vehicle per annum and a density of one vehicle per hour; with zero traffic there are no variable costs. The surface will then rise with an increase in volume and with an increase in density. Corresponding to this average variable cost surface could be constructed a marginal cost surface which would begin at the same point and increase with volume and density, but rising faster in each direction than the average variable cost surface.

The average fixed cost surface, being entirely a function of volume, will have the cross-section shown in diagram 1*a* at any density. Vertical addition of the average variable cost surface and the average fixed cost surface gives the average total cost surface.

No attempt is made to draw these three-dimensional surfaces as our limitation to two-dimensional paper precludes their use for geometric analysis. The cost picture has been developed in terms of three variables, however, for purposes of clarity and accuracy. Use of mathematical techniques depicting the surfaces in terms of functions in three variables would enable us to proceed with the analysis in this way. The purpose of this book is to develop the principles of analysis of highway problems, and these are more clearly understood if presented in terms of geometry rather than calculus. We shall therefore reduce the three variables of costs to two at the expense of some accuracy, in order to present more clearly the fundamental principles of the analysis.

The Simplification of Costs

In reducing the three variables (cost, volume, and density) to two we must retain cost as a separate variable and are therefore forced to find a single measure of traffic to replace the two measures, volume and density. Volume measures the quantity of traffic over a period of time while density measures rate of flow at a given time. This is analogous to the distinction between mileage and speed of a vehicle. When we say a vehicle travels 12,000 miles per year we refer to its total mileage over a period of time. But when we say it is travelling at 60 miles per hour we refer to its speed at an instant, not its mileage during one hour. To express 12,000 miles per year as 1,000 miles per month is not saying the same thing, since the vehicle might not actually travel 1,000 miles in any single month. We are thus expressing the average rather than the actual monthly mileage. But if we express 60 miles per hour as one mile per minute or 88 feet per second, or 1,440 miles per day, or 525,600 miles per year, we are expressing exactly the same idea. Since traffic volume is analogous to annual mileage, and density to speed, we cannot change the time period of volume measurement without involving averages, but we can change the time period of the density measurement. Both the measures of volume and density could therefore be expressed in vehicle units per annum, and it might be thought that this solves our problem. It does not do so, however, and we mention it only to illustrate the dangers of assuming the simple answer to be correct. It fails because although using the same measure, vehicles per annum, we are still measuring two concepts, volume and density. To overlook this and plot a two-dimensional diagram in this way would make the tacit assumption that at any time the density corresponds to the average density of the annual volume. This would merely take an oblique cross-section of our three-dimensional surface, and the assumption of a constant density day and night is about the most unrealistic we could make. The fallacy is analogous to saying that since this car is travelling at 60 miles per hour it will travel 525,600 miles in a year. In practice speed varies, and so does traffic density.

The problem, therefore, is not merely to reduce to one our two units of measurement but to reduce to one the two concepts which we are trying to measure. The technique of doing this can again be illustrated by analogy with the speed and mileage of a vehicle.

Sixty miles per hour is the same speed as 88 feet per second, but if a car actually travels 88 feet in one second this is a measure of distance covered or average speed during one second, just as 12,000 miles per annum is an average speed of 1.37 miles per hour. But whereas the average speed over one year gives a very poor idea of the actual speed of the vehicle at any time, the average speed over one second is a close approximation to the actual speed. In other words, the shorter the time period the closer the approximation between mileage and speed. Similarly traffic volume is a measure of average density, and the shorter the time period over which we measure volume the closer will that measure of volume approximate to a measure of density. If we select an appropriate time period for our measurement of volume we can therefore regard it as a measurement of density. The inaccuracy involved is the price of simplifying three dimensions to two.

In selecting the appropriate time period we are faced with a dilemma. The shorter the time period the smaller the inaccuracy involved in regarding volume as a measure of density. But if we shorten the time period by averaging volume this measure loses its significance, just as 12,000 miles in one year is not the same thing as 1,000 miles in one month, and has even less significance as 1.37 miles in one hour. While averages involve inaccuracy, however, it is just as meaningful to measure actual mileage over one hour as it is over one year. This raises the question of why we select one year as the period over which to measure traffic volume. This is entirely a matter of convenience. One year is long enough to avoid hourly, daily, and seasonal variation and a convenient period for the amortization of capital costs, but it is an entirely arbitrary choice. There is no reason why if we can divide fixed costs into equal annual costs and average those over annual traffic volume, we cannot do the same on an hourly basis. Similarly variable highway costs which are a function of traffic volume could be costed just as easily over one month or one hour as one year. The difficulty with the shorter time period is that there are many more periods over the life of a highway. If the anticipated life of a highway is twenty-five years and we use annual periods for analysis there are twenty-five periods to analyse. If we use an hourly time basis there are 219,150, and these will show variations up and down in traffic volume with hourly and seasonal peaks, while the annual periods

Diagram 2

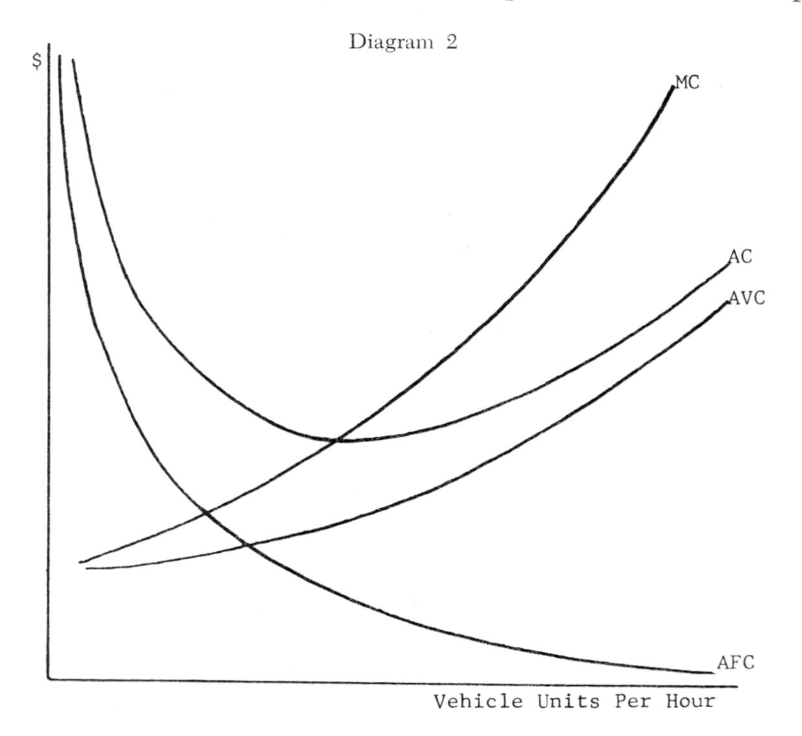

would show a continuous rise in volume with secular growth. Since our demand pattern shows such variations, however, as we shall see in the next chapter, and account must be taken of these, there is no reason why the cost analysis should not be equally voluminous. As has already been stated, use of mathematical techniques which can handle numerous variables would considerably simplify the practical application of this theory. That the practical use of two-dimensional geometry on hourly intervals would involve nearly a quarter of a million diagrams does not prevent us from using it to illustrate simply the principles of analysis.

By using vehicles per hour rather than vehicles per annum as the measure of volume we can avoid the problem of meaningless averages and give rise to no theoretical obstacles. The only difficulties are practical ones which can be overcome by use of mathematics. At the same time the number of vehicles over one hour is a sufficiently close approximation to traffic density during that hour to satisfy our purposes. If the reader is worried by variations in density during one hour he can think in terms of vehicles per minute. The principles will be just the same.

We have now reduced the two variables volume and density to one variable. This is volume over one hour which can also measure average density with sufficient approximation to actual density. We can therefore depict the cost curves on a two-dimensional diagram as in diagram 2. This shows great similarity to the cost curves of any firm and will be used in much the same way in chapter IV. We must first examine the demand functions with which these costs must be reconciled and this is the task of the next chapter.

CHAPTER THREE

THE DEMAND FOR HIGHWAY TRANSPORTATION

The Meaning of Demand

The demand for highway transportation is the same concept as that of the demand for any good or service. It is a series of the amounts of the service which consumers are willing to purchase at a series of prices in a given period of time. Demand must be clearly distinguished from desires, wants, and needs which are not backed by a willingness and ability to pay. It therefore depends on the two forces of tastes and preferences of consumers, and income distribution.

The price is the amount which the consumer pays for a good or service per unit consumed. In the case of highway transportation it takes the form of all the costs directly borne by the consumer, vehicle costs, users' personal costs, and fuel taxes. The significance of the demand curve is in showing not only how many vehicle units will travel at a certain price, but how many units consider the value of a trip to be equal to or greater than a certain amount. Thus, while in the analysis of costs the money axis showed the cost of highway transportation in relation to traffic volume and density, the demand curve shows the value of transportation to the consumer, as it is expressed in the form of willingness to purchase a certain amount at a certain price. While we may refer to the demand curve as showing the relationship between traffic volume and cost, since the money axis of our diagrams measures costs and the actual payment is in the form of operating costs, the curve should be thought of as showing willingness and ability to bear a certain cost; and it is in this sense that the curve is used in the analysis of the following chapter.

In this analysis we restrict demand to that of vehicular travel, any other desire for highway improvements for any other reason, such as unemployment relief, being regarded as a negative community cost. In this way we are able to isolate the primary function of highways from the host of other issues, some relevant and some irrelevant, with which highway planning is invariably associated in practice.

One of the difficulties of analysing highway transportation as a service is that it is not homogeneous. It is meaningless to speak of vehicle miles without reference to location. The utility of transportation comes not from consuming so many miles of it but in getting from point *A* to point *B*. Whenever we speak of demand therefore it must have reference to a particular journey. Just as we analysed costs with reference to a specific project, so we must analyse demand for travel over a specific section of road.

Definition of the unit of highway transportation involves the same difficulty that we faced in the analysis of variable costs, the heterogeneity of traffic. We overcome it by use of the standard vehicle unit defined in the previous chapter. The unit of highway transportation is the journey by one vehicle unit over the specified section of road. Whether we separate demand for travel in each direction depends on the nature of the project under analysis. This refinement might be worth while in the analysis of divided highways, but not of two-lane roads. The problems involved were examined in the previous chapter on costs, and the same decision must be made for demand analysis as for cost analysis. For some projects both costs and demand will be analysed separately for each direction of travel, in others the opposite flows will be combined, but the analysis of a single project must be consistent for both costs and demand.

Any statement of demand must have reference to a period of time, for a certain number of units per week is quite a different thing from the same number per day. In the case of highway transportation, where demand varies widely with peaks and growth and the non-storable nature of the service precludes the use of inventories to even out an irregular demand stream, we must define not only the duration of the period, but the specific period of that duration. We must state demand not only in terms of so many vehicle units per hour at a certain price, but also state whether the hour in question is a peak hour, mid-day hour, or night hour, whether it is a week-day or Sunday, summer or winter.

The choice of an appropriate unit of time involves many of the same problems that were discussed in the last chapter on costs. We noted that traffic could be expressed in terms of both volume and density, and suggested that vehicle units per hour was an appropriate meassure of volume which would also serve as an approximate measure of density during the hour. The same arguments apply to demand, which manifests itself as a stream of traffic of irregular density over time. We can again use vehicle units per hour as a measurement of the volume of demand which approximates closely to the extent of demand during the hour. This expresses demand at a moment of time as a potential rate of flow, and this rate of flow will vary with daily, weekly, and seasonal peaks, and with growth over time. At any moment it will show a series of inversely related levels of cost and volume. Before expressing demand as a demand curve for use in the geometric analysis we must examine the factors which determine the extent of demand.

The Components of Demand

In estimating the extent of demand it is convenient to examine separately four sources of traffic; existing traffic, diverted traffic, generated traffic, and secular growth.

Where the project consists of an improvement in or replacement of an existing road, the volume of traffic previously using the road will continue to do so at the same or lower cost. Since the object of the improvement is presumably to reduce cost, how much of this traffic would continue to flow at higher cost is of little or no significance. The extent of demand by existing traffic is easily estimated by a traffic survey before the improvement. If the project involves building a completely new road there will be no existing traffic.

Where a new road is built or when the cost level on an existing road is reduced by improvement, traffic might be diverted from other roads. The extent of diversion will depend on the layout of the road system in relation to patterns of movement. The closer together highways are over an area, the greater will be the tendency for traffic to divert to the new road. The catchment area for the new road or improvement will radiate with diminishing intensity because the time and distance involved in detours from a distant route to use the new facility will offset the potential reductions in operating cost. In a given situation the amount of diversion will be greater the lower the level of cost on the new road and the demand curve of this group of traffic will therefore be downward-sloping.

Further traffic will be generated by the existence of the new highway or improvement. This will be traffic for which the utility of movement was below the previous operating cost, but above the operating cost on the new road. Some of this traffic may have previously used alternative forms of transport, but much of it will be completely new. Where a new highway opens up an area for economic development, generated traffic so caused might well become a large proportion of the total. Again we find that the lower the level of cost on the new road the greater will be the volume of generated traffic and the demand curve of this group will therefore be downward-sloping.

Finally, account must be taken of growth, or decline, in traffic volume over time. Past trends of growth in traffic volumes must be projected into the future to estimate this; and we must bear in mind that a major road improvement might increase the rate of growth in economic activity and thereby foster growth in traffic volume. The distinction between generated traffic and secular growth is not precise, but is of little significance since it is with total demand that we are concerned. The breakdown of demand into the above categories is purely for the convenience of estimating total demand which is the sum of the four components. Further discussion of the techniques of estimating demand is deferred until chapter v.

The Demand Curve

The demand curve expresses a relationship between three variables; vehicle units per hour, cost of the trip, and time. The first presents no further difficulties, but cost measured in dollars involves the problem of a changing value of money over the life of the highway, which is the period for which demand must be forecast. Changes in the value of money cause changes in the real value of the cost axis. The same problem applies to the analysis of costs. We can avoid the necessity for detailed forecasts of future changes in the value of money by the simplifying assumption that it will vary uniformly over the items with which we are concerned—the value of travel, vehicle costs, values to be placed on time and accidents, and so on. All the curves used in our analysis will then have the same relationship to each other whatever changes take place in the real value of money. Time is an important variable in that the level of demand and its elasticity will vary with peaks and growth. A complete representation of the demand curve would therefore involve a time axis on a scale sufficient to show daily peaks clearly but stretching over the complete life of the highway. Such detail is, however, unnecessary for exposition of the technique of analysis, and is of doubtful practical value in view of the considerable error margins in forecast data. The demand curve will be effectively the same for all week-days of a certain season of a given year and there is therefore little point in showing each day separately. The variations between week-day and week-end traffic will be different for different projects. In some cases it might be worth analysing the Sunday pattern separately from the week-day, in others they can be averaged with no great loss of accuracy. In a few cases it might even be worth showing the demand on public holidays separately. Similarly the extent of seasonal variations will differ. In some cases separate curves for each month might be worth while; in others two curves showing summer and winter will suffice; while in still others seasonal variations will be so slight that a single annual average can be used.

The degree of detail used in the analysis can vary from a single curve showing the average daily peak pattern for a whole year to a separate pattern for each day of the year. The degree of detail practicable in the analysis depends on the methods used. Great detail would make the geometric technique extremely cumbersome but mathematics could conveniently handle considerable detail (the possibilities of using computers are virtually unlimited). To keep the geometric exposition of the technique of analysis clear and simple we shall use a single average daily pattern for the year, but the method of introducing greater detail will be made clear.

Although the use of an annual average daily traffic pattern considerably simplifies the analysis, it forces us to use four variables instead of three. Time, a continuous variable, is broken down into time of day and the year in question. Thus, while our three-dimensional demand surface now involves a time axis only twenty-four hours long, instead of maybe twenty-five years long, we need a separate surface for each of the twenty-five years. If we require separate week-day and Sunday patterns for each of two seasons there would be four surfaces for each year or a total of a hundred surfaces, and so on. Twenty-five years was suggested earlier as the maximum anticipation of highway life, and we must forecast demand for the whole life in order to assess the efficiency of a project. The difficulties involved in long-range traffic forecasting are so great, however, that most authorities limit themselves to fifteen or twenty years at most. This discrepancy must be overcome by an appropriate choice of anticipated life. If demand cannot be forecast more than twenty years into the future, then that is the maximum anticipated life of the highway. If it is confidently expected that the highway will be useful for twenty-five, then some forecast of the last five years' traffic is implied. Whatever life is chosen we must forecast demand, from which comes utility, over the same period that is used for the amortization of costs.

A typical set of demand surfaces is shown in diagram 3, where the surfaces have been drawn at five-year intervals. The curve *Yr.* 1 in the upper part of the diagram shows the traffic volume as a function of cost at 6.00 p.m. on an average day in the first year. The corresponding curve *Yr. 1* in the lower diagram shows the traffic volume at each hour of the average day at a cost level $ *Ox*. These two curves are merely cross-sections of one surface; similar cross-sections could show the demand curve at other hours of the day, and the peak pattern at other cost levels. The curves labelled *Yr. 6*, *Yr. 11*, *Yr. 16*, and *Yr. 21* similarly show cross sections of the surfaces for the sixth, eleventh, sixteenth, and twenty-first years. A set of surfaces such as this would represent a fairly steady rate of traffic growth over the years with much the same daily peak pattern.

The curves used in the following analysis, where we are limited to two-dimensional geometry, are of the type shown in the upper part of the diagram—single demand curves each representing the demand relationship between traffic volume and cost at a given time of day in a given year.

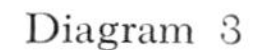

Diagram 3

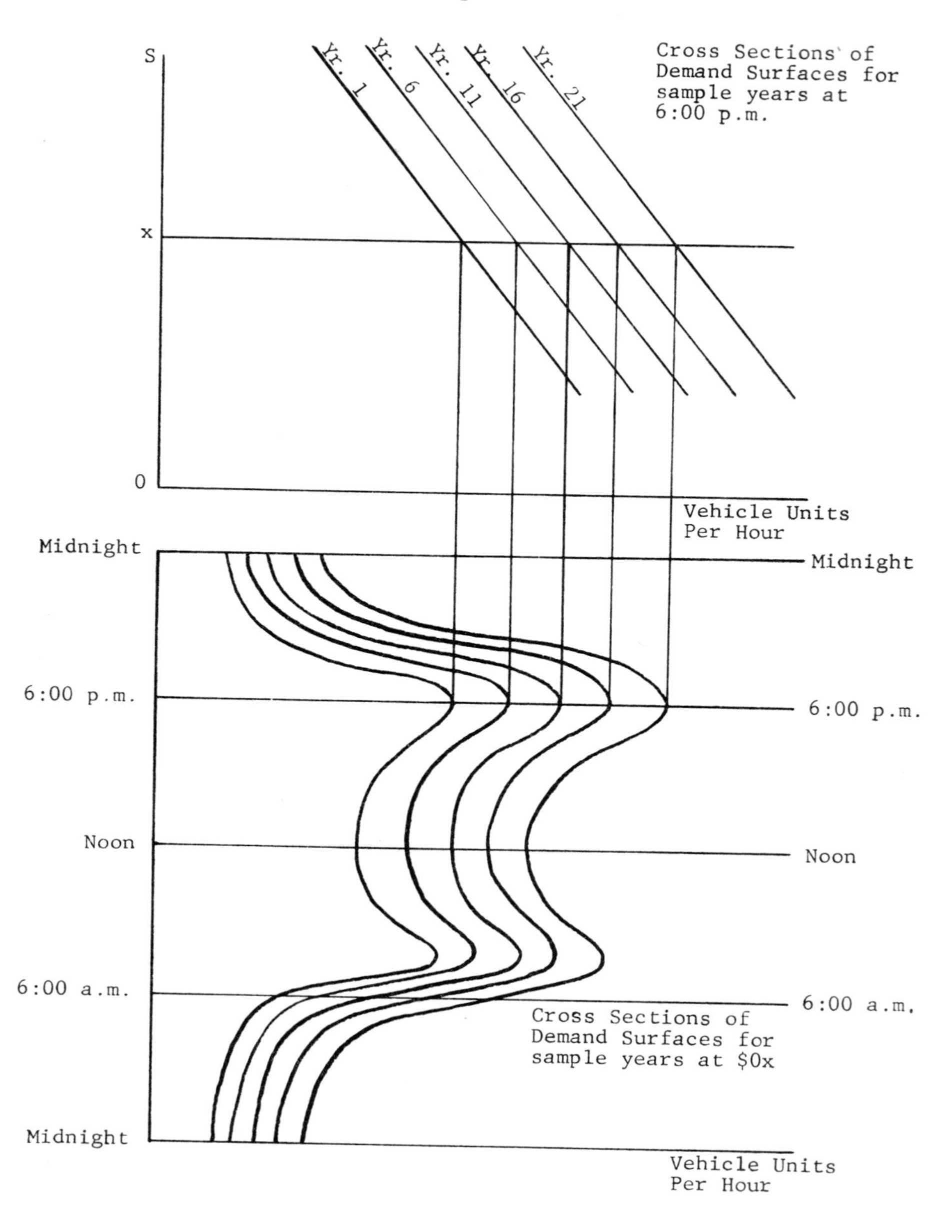

CHAPTER FOUR

THE OPTIMUM SOLUTION

IN THE LAST TWO CHAPTERS we have analysed the forces to be reconciled in highway planning from the cost and demand sides. Now comes the task of reconciliation to determine what solution to any problem gives the correct balance between the various forces. The correct solution will be the optimum one, in the sense that any other solution would involve either costs which are not justified by yielding benefits greater than the cost involved, or an opportunity to incur costs which would yield benefits greater than the costs involved. Throughout this chapter it is assumed that we have complete and accurate data of all the relevant cost and demand schedules; problems of data collection and estimation are discussed in the next chapter.

In practice it will rarely be possible to achieve perfection, and to some extent therefore a theoretical model which precisely defines the optimum solution will be unrealistic. However, the best solution in practice is the closest approximation to the theoretical optimum. To compare practical solutions with this in mind we need some way of assessing how closely they approximate to perfection. The best yardstick is a precise formulation of what constitutes an optimum solution, and the aim of this chapter is to provide that yardstick.

The impossibility of achieving perfection in practice leads to a dilemma in the formulation of the theory, however. The best plan for a road depends on how much traffic will use it, but as we have seen in our analysis of demand the amount of traffic is dependent on the cost of transportation, which is in turn dependent on the pricing or taxation system. Given the demand conditions, there will be an optimum volume of traffic for any given plan of the road. The best possible plan will be that which yields the greatest excess of benefit over cost when carrying its optimum volume of traffic. It is the function of the pricing system to endeavour to encourage an optimum traffic flow, but to do this for all roads at all times would call for a pricing system the complexity of which would be neither practicable nor justifiable in view of the high administrative costs. In practice, the pricing system must be a compromise solution which comes as close as possible to achieving optimum traffic volumes while retaining simplicity. But on some roads at some times this means that the traffic volume we actually have might be considerably different from what would be the optimum. This poses the problem of whether the criterion for selecting the best plan for a road should be the net benefit derived when an optimum volume of traffic uses the road, or the net benefit when the traffic volume is that which would actually result from the pricing system. The results of the analysis in some cases might well hinge on this decision.

Fortunately, the technique of analysis will be the same in either case and each has its value. Planning on the basis of optimum traffic volumes guides us in the objectives of the pricing system, and provides the data on which the pricing system is based. Once the pricing system is determined, however, we must plan projects on the basis of the traffic volumes which will actually flow. The first part of this chapter develops the analysis on the assumption that traffic volumes will be optimum. Consideration is then given to the modifications necessary to take account of actual rather than ideal conditions. The results of the first part are used to determine the pricing system, which is considered in chapters VIII and IX.

Determination of the Optimum Traffic Volume

The optimum volume of traffic on a given road at a given time depends on the relationship between the appropriate demand curve and cost curves. On diagram 4 we have the marginal cost curve for the road in question, as developed

Diagram 4

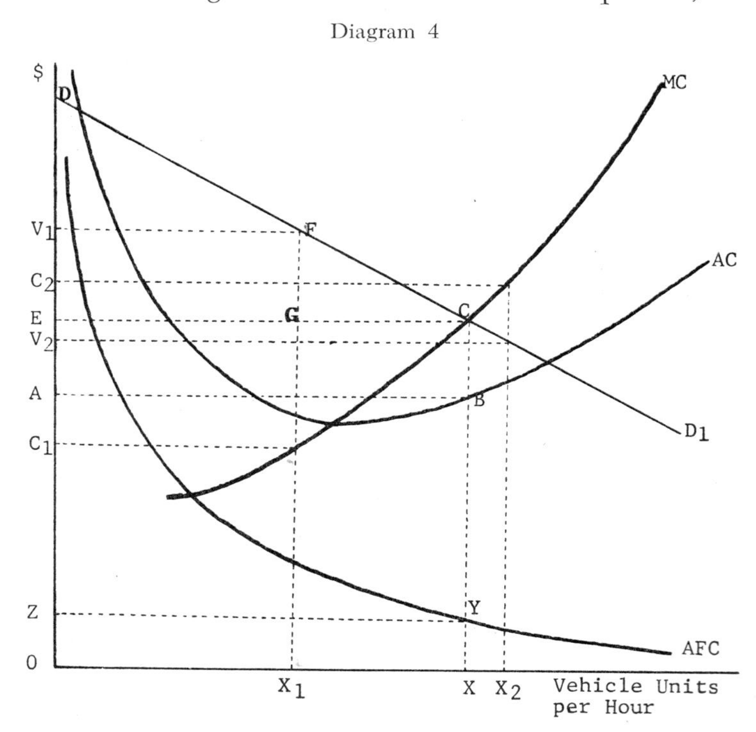

in chapter II, and the demand curve for travel over this road at the appropriate time, as developed in chapter III.

If the traffic volume is OX_1, with cost per trip to the user OV_1, the marginal cost is OC_1. If price were reduced slightly below OV_1, more traffic would flow and we can therefore say that the value of the $(X_1 + 1)$th trip is slightly below OV_1, since at this price someone would just consider the trip worth-while. However, if price were slightly reduced so as to allow this trip, the additional cost would be slightly above OC_1. The marginal cost curve, it must be remembered, takes account of all the additional costs of another trip. In this case the value of the $(X_1 + 1)$th trip is greater than the additional costs involved and this trip is therefore worth while. The optimum volume of traffic must therefore be greater than OX_1 since if we had this volume it would be worth having more. This argument would apply to any volume for which the demand curve is above the marginal cost curve.

However, if we had traffic volume OX_2 at price OV_2 the marginal cost would be OC_2. If price were raised slightly above OV_2 traffic volume would fall, showing that the value of the X_2th trip to the user is just OV_2. However, as a result of this trip being made additional costs of OC_2 were incurred. Thus we can see that since OC_2 is greater than OV_2 this last trip was not worth its cost and traffic volume should be reduced. This argument applies to any volume for which the marginal cost curve lies above the demand curve.

Since the marginal cost curve is characteristically upward-sloping and the demand curve downward-sloping there will be some point at which the two intersect, unless one lies wholly above the other. It is impossible for the demand curve to lie wholly above the marginal cost curve, since the latter rises to infinity at that volume at which traffic reaches a standstill through congestion. If the demand curve lies wholly below the marginal cost curve the optimum traffic volume is zero since even the first unit is not worth its cost. That volume represented by the point of intersection of the demand curve and marginal cost curve is clearly the optimum, since at any lower volume the demand curve lies above the marginal cost curve, showing a greater volume to be worth while, while at any volume greater than that represented by the point of intersection the marginal cost curve lies above the demand curve, indicating that a smaller traffic volume is desirable.

We saw in chapter III that the demand curve will be different at different times of the day and seasons of the year. Similarly, if we are to be precise, the marginal cost curve might be different at different times. Two of the components of marginal cost are vehicle operating costs and the users's personal costs of time, inconvenience, and risk. Vehicle operating costs might well be higher in winter than summer, and the costs of fatigue and risk higher at night than in daylight. Since the two curves, particularly the demand curve, are not constant over time, neither will the optimum volume of traffic be the same at different times of day or seasons of the year. Thus the optimum traffic volume must be specified not only in relation to a particular plan for the highway but with reference to a particular time.

The Measurement of Benefit

It has been shown that for the cost conditions depicted in diagram 4, the optimum volume of traffic is *OX*, which would flow at a cost per trip to the user of *OE*. This traffic must therefore yield benefits to the users in total of at least the amount they would be prepared to pay, i.e., *OECX*. But some of these users value the trip at an amount greater than *OE* as is shown by their willingness to travel at a higher price; OX_1 units for example would travel at a price of OV_1. The problem of expressing the monetary value of the total benefit derived is one to which economists have devoted a great deal of thought, without reaching agreement.[1] The simple technique is to measure total benefit as the area *ODCX*, but three objections are raised to this.

The first is that even if we are dealing with an individual's demand curve and a small price fall from say OV_1 to OE we cannot say that the area V_1FCE measures the increment in net benefit to the consumer.[2] The author does not accept this objection for the present use of the concept of benefit, and even those who object agree that the area in question is at least a very close approximation to the increment in benefit.

The second objection is that even if we accept the area V_1FCE as a measure of net benefit for a small price change we cannot apply the same technique to the whole triangle *EDC*. This is largely because the upper end of the demand curve is entirely conjectural. While this objection is true for any practical application of the concept of benefit, it fortunately is not serious for our present purposes, for we are primarily interested in comparing the benefits which will accrue from alternative plans. The arbitrariness of the assumed position of the demand curve will affect all plans equally, and the results of our analysis would be the same whether we assume the shape of the demand curve at its upper end to be *DF* or V_1F or any other similar shape, providing of course that we make the same assumption for each plan under consideration. Thus, while the actual monetary measure of benefit must not be considered as absolutely precise in any one case when we base it on an arbitrarily assumed shape of the demand curve, the comparison between alternatives will still be valid. If we overestimate or underestimate benefit from each plan by the same amount, we still know which yields the most benefit; and this is our primary interest.

The third objection is the most serious to the whole theory of welfare economics. It is that one cannot aggregate the benefit derived by different people and arrive at a meaningful total for the whole community. This, of course, is what we are doing in measuring the area under the market demand curve. Valid though this objection is it remains true that unless we make some assumption about interpersonal comparisons, economics can offer no help in problems of policy such as that of highway planning. Our assumption is simply that if one person derives a benefit of $10 and another of $15 between them they are $25 better off; and that this situation is preferable to one which would make two other people $20 better off between them. It cannot be proven that from the standpoint of the community as a whole it is better to make one group of people $25 better off rather than another group $20 better off, since community welfare depends on the distribution of wealth as well as its total. It might well be better to have a smaller cake fairly divided than a larger cake unfairly divided. The theory of this chapter is based on the assumption that the existing distribution of income is roughly ideal, which can be defended on the ground that since society has modified the free market income distribution to a large extent already by such means as progressive taxation and transfer payments, it has presumably now reached the stage where there is no consensus of opinion on the desirability of further revision. Were it generally agreed that changes in the existing distribution would be desirable these would be brought about by normal budgetary procedures. There is certainly no reason for using the planning and financing of one particular economic activity, such as highway construction, to redress any remaining inequities in the distribution of income, especially when there is no way of knowing whether, and if so where, such inequities exist. The distribution of the net benefit derived from highway construction, as distinct from the redistribution of existing wealth and income, resorts to the financial problem of how much each person should pay in user taxes or receive by way of compensation, in relation to the benefit which he derives from the highways. To this problem we turn our attention in later chapters.

Thus despite the objections, but with the reservation that the results of our analysis cannot be proven to be precisely accurate and must be interpreted with caution, we shall measure benefit by the area under the demand curve on the grounds that this method offers the best guidance available in problems of practical policy, that in all probability the solution achieved will be the best attainable, and that there is certainly no way of showing, nor even any reason to believe, that any other solution would be preferable. The benefit derived in one hour from *OX* vehicle units travelling over the road in question is therefore the area *ODCX*.

The same technique of measuring benefit can be applied to more complex cases where sections of road cannot be considered in isolation. This will normally be the case where any extensive improvement or construction programme is under consideration, since the demand for travel over one section of road will be dependent in part on cost levels on other sections of road. The principles of analysing such a case can be shown by an example using the same geometric methods, though some of the complex cases encountered in practice might be more conveniently handled by employing the same principles, but using the medium of calculus.

Diagram 5 shows the alternative locations for a road between points *X* and *Y*, which points are already determined by topography. Between them lies the city *Z*, which poses the choice between a road through the city, a road bypassing the city, or both. Plan 1 would comprise sections *A* and *B*; plan 2 sections *C*,

Diagram 5

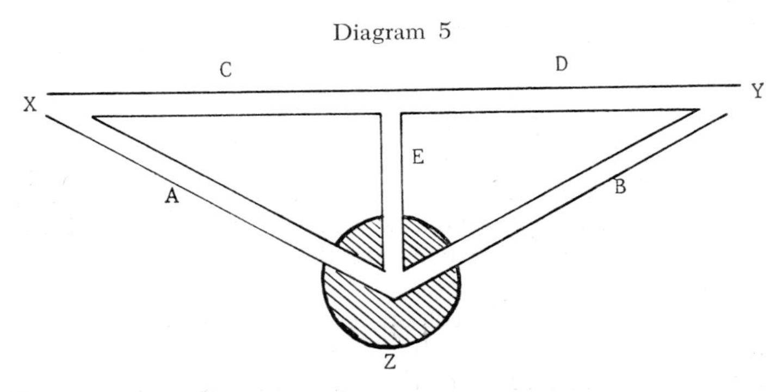

D, and *E*; and plan 3 sections *A*, *B*, *C*, and *D*. There will be three separate traffic flows to consider, *X-Y*, *X-Z*, and *Y-Z*, for each of which there will be a demand pattern. The problem is to determine the optimum traffic volumes and net benefit for each of the three plans.

Plan 1 involves consideration of the three demand curves and the cost curves for sections *A* and *B*. Diagram 6 shows the marginal cost curve for section *A* and the demand curve for *X-Z* traffic. To this demand must be added the *X-Y* traffic flowing over section *A* to arrive at the optimum traffic volume and level of marginal cost. Given the *X-Z* demand curve, and assuming that an optimum traffic volume is attained, the marginal cost on section *A* will be a function of the volume of *X-Y* traffic. This relationship is easily derived from diagram 6, and is shown on diagram 7. If there is no *X-Y* traffic the level of marginal cost

Diagram 6

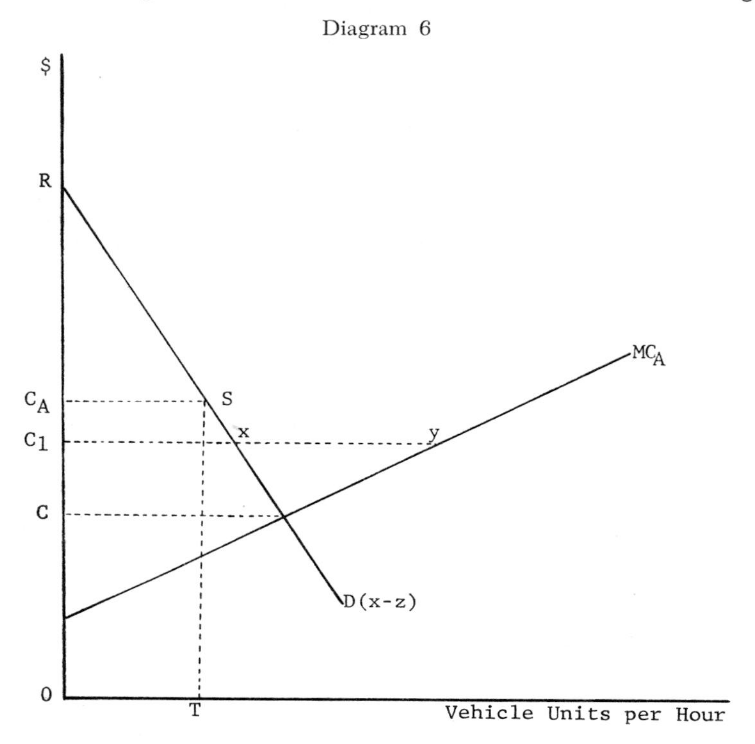

on section *A* is *OC*. If the volume of *X-Y* traffic is *xy* this traffic must be added to the demand curve for *X-Z* traffic to give a total demand curve which will intersect the MC_A curve at *y*, giving marginal cost OC_1. In this way the curve showing MC_A as a function of the volume of *X-Y* traffic is built up, the marginal cost in diagram 7 for any volume of *X-Y* traffic being that amount at which the MC_A curve in diagram 6 lies to the right of the *X-Z* demand curve by the volume of *X-Y* traffic.

In the same way the marginal cost on section *B* can be shown as a function of the volume of *X-Y* traffic on section *B*, using the marginal cost curve for *B* and the *Y-Z* demand curve instead of the marginal cost curve for *A* and the *X-Z* demand curve in diagram 6. On diagram 7 the two marginal cost curves as functions of the volume of *X-Y* traffic are added vertically to show the marginal cost on the complete trip *X-Y* as a function of *X-Y* traffic, and the demand curve for *X-Y* traffic is superimposed on this. The point of intersection between these shows the optimum volume of *X-Y* traffic, which in turn shows the optimum levels of marginal cost C_A on section *A* and C_B on section *B*. The optimum volume of *X-Z* traffic can now be read from the demand curve in diagram 6 at the marginal cost level C_A, and the optimum volume of *Y-Z* traffic is similarly read from a diagram similar to diagram 6 but depicting section *B*.

The benefit from this plan in the hour represented by the curves used is the sum of the benefits to the three traffic flows. *X-Z* traffic derives benefit *ORST* on diagram 6, *Y-Z* traffic the similar area on the corresponding diagram for section *B*, and *X-Y* traffic the area *ORST* on diagram 7.

Diagram 7

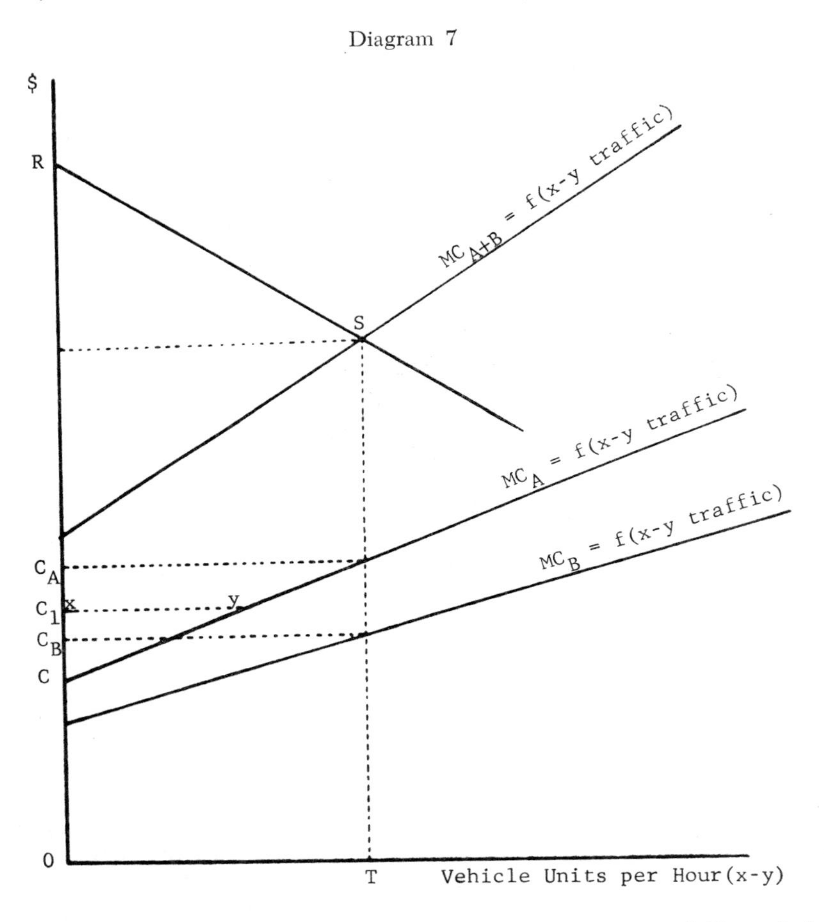

The analysis of plan 2, comprising the bypass sections *C* and *D*, and the link road *E*, is made more complicated by the fact that each traffic stream uses two sections of road, the *X-Z* traffic using sections *C* and *E*, the *Y-Z* traffic *D* and *E*, and the *X-Y* traffic *C* and *D*. We begin by assuming a certain level of marginal cost *OE* to exist on section *E*. The *X-Z* demand curve is now lowered vertically by the amount *OE* to give the demand curve for *X-Z* traffic for travel over section *C*. Similarly the *Y-Z* demand curve is lowered by *OE* to give the demand curve for *Y-Z* traffic for travel over section *D*. We can now analyse sections *C* and *D* in exactly the same way that sections *A* and *B* were analysed for plan 1, and arrive at the optimum volumes of *X-Z* and *Y-Z* traffic, still on the assumption that marginal cost on section *E* is *OE*. The sum of the *X-Z* and *Y-Z* traffic volumes is the traffic on section *E* and this volume *M* is plotted against cost level OE on diagram 8, on which is also drawn the marginal cost curve for section E. It is now seen that at the assumed level of marginal cost on section E, the traffic volume is OM, but at this volume the marginal cost is OC. The assumed level of marginal cost was therefore too high and we can repeat the

Diagram 8

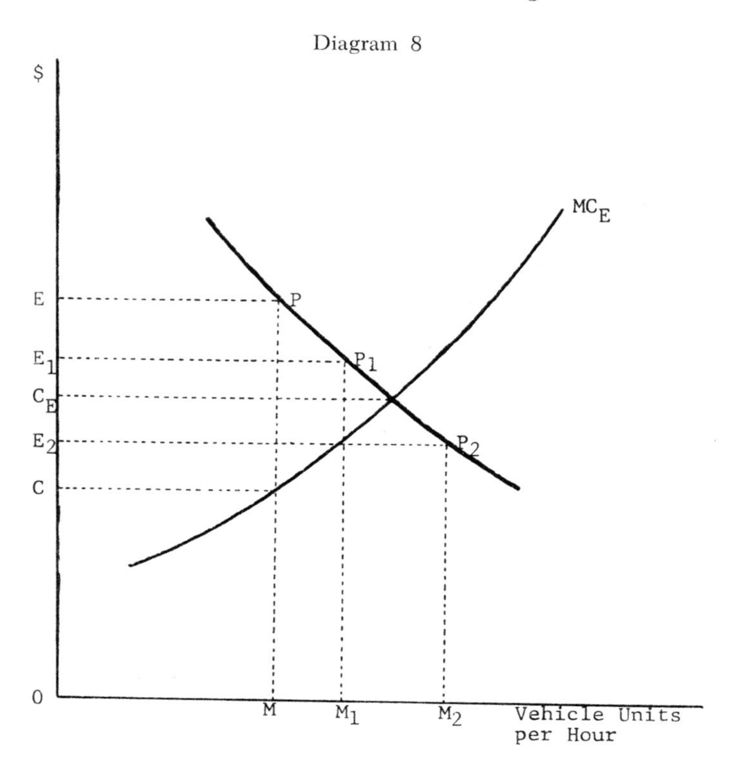

analysis on the assumption of a lower level of marginal cost OE_1, which results in traffic volume OM_1 on section E. Similarly, the assumption of marginal cost-level OE_2 would result in traffic volume OM_2. A series of such assumptions enables us to trace the curve P, P_1, P_2, which shows the volume of traffic which would flow on section E as a function of the *assumed* level of marginal cost on that section. The marginal cost curve for section E, however, shows the *actual* level of marginal cost as a function of the volume of traffic. The only assumed level of marginal cost which can be accurate is that represented by the point of intersection of these two curves. This shows the correct assumption for the marginal cost level on section E to be OC_E. Basing the analysis on this assumption will yield the correct optimum traffic volumes for the three routes X-Z, Y-Z, and X-Y. Once these are known, benefit can be read from the areas under the relevant demand curves.

The trial-and-error method of analysing this case, and the even more numerous stages which would be necessary for more complicated plans and combinations of traffic flows, result from the limitations of geometric analysis where numerous variables are involved. These difficulties can be overcome by the use of mathematics, and in practice much of the detailed calculation can be delegated to computers.

Plan 3, which encompasses sections *A*, *B*, *C*, and *D*, is by far the most simple to analyse since no combined traffic flows are involved. The *X-Z* traffic uses section *A*, the *Y-Z* traffic section *B*, and the *X-Y* traffic sections *C* and *D*. There is no need for the link road *E*. The three routes can therefore be analysed separately by the simple technique outlined above for a single road and the results totalled.

THE CONCEPT OF NET BENEFIT

So far we have concentrated on measuring the benefit derived by an optimum volume of traffic using the highway under analysis in the course of one hour. The important criterion for planning purposes, however, is the amount by which benefits exceed costs. To measure this net benefit for one hour in the case of a completely new highway we must deduct the total costs attributable to traffic in that hour from the total benefit. In diagram 4 the total costs are represented by the area *OABX*, being the average cost per vehicle unit multiplied by the number of units. The net benefit over and above all costs is the area *ABCD*. In the case of the more complex plans shown in diagram 5 the same technique is used. The analysis outlined above determined the optimum volumes of *X-Z*, *Y-Z*, and *X-Y* traffic and total benefit is the sum of the areas under these demand curves. Once we have determined these volumes, the optimum traffic volume on each section of road is known. The total cost for each section is the rectangular area under the average cost curve for that section of road at the indicated traffic volume. Where a plan involves more than one section of road, total cost for the plan is the sum of the costs for each section. By deducting total cost for the plan from total benefit we determine the net benefit derived in the hour in question.

The amount of net benefit derived in one hour can be multiplied by the number of hours in the first year for which the demand curves used in the calculation are typical. Similar calculations can be made for other hours and the results totalled to give the net benefit which would result in the first year of operation from the construction of the road in question. How many demand curves we use would depend on the degree of accuracy required. The extreme would be to use as many curves as there are hours in the year, but in view of the fact that we are working with estimated data subject to error margins it will probably suffice to take a representative curve, or set of curves, for week-day off-peak daylight traffic, one for peak traffic, and one for night traffic; with separate sets for Saturdays, Sundays, and public holidays. Where seasonal variations are significant different curves may be used for summer and winter.

Having arrived at the net benefit which the plan would yield in the first year, we can make similar calculations for the anticipated demand conditions in each of the other years of the anticipated life of the highway. This stream of potential net benefits can then be discounted at the assumed rate of interest to arrive at a single value for the net benefit which would result over its anticipated life from the plan under analysis.

The concept of net benefit is equally applicable in the analysis of an existing highway and is of importance where replacement is under consideration. The optimum volume of traffic on an existing highway is shown by the point of intersection of the marginal cost and the demand curves in the same way as for a proposed new highway, and total benefit in similarly measured by the area under the demand curve. But the costs which should be deducted from benefit to arrive at net benefit must exclude historic costs in the case of an existing highway, whereas these were included for a proposed new highway. The once-for-all capital expenditures necessary for the construction of a new highway are important considerations for the analysis of planning until those expenditures are made. Once they are made, they are not redeemable and cease to be a relevant cost of use of the highway. This does not mean that amortization of capital expenditures should not be paid from vehicle taxation; it means only that the extent of this capital cost is not affected by the extent of use of the road, or whether it is used at all. How far amortization should actually be included in determining the level of vehicle taxation is a financial problem which will be examined in later chapters.

In the planning analysis we are essentially concerned with decisions about the future. Potential future benefits and costs are important, but irredeemable past expenditures are not. Thus the costs which must be deducted from benefits to arrive at net benefit in the case of an existing highway exclude any outstanding portion of right-of-way costs, development costs, and construction costs, but include all future costs of continued use of the highway, which include fixed maintenance costs, administration costs, and all variable highway, community, vehicle, and users' personal costs. It is not necessary to discount the potential future stream of net benefits from an existing highway to arrive at a single figure, since they are more useful in the planning analysis in the form of a future stream. No decision as to the anticipated future life of an existing highway is therefore necessary.

Where the plan under consideration is for a new highway to replace an existing highway, the net benefit from the project will be the anticipated additional benefits minus the anticipated additional costs. The net benefit from the new highway as determined above will not all be additional benefit, since the existing highway would have continued to yield some net benefit had the new highway not replaced it. The net benefit from a replacement project is therefore the discounted stream of the excess of net benefit from the new highway over the net benefit which would have been derived from the old highway. Thus if we symbolize benefit and total cost of the new highway as B_1 and C_1 in the first year, B_2 and C_2 in the second year, etc., and benefit and costs other than historic costs on the existing highway as b_1 and c_1 in the first year, etc., the stream of net benefit anticipated from the replacement will be $(B_1{-}C_1) - (b_1{-}c_1)$; $(B_2{-}C_2) - (b_2{-}c_2)$, etc.

We are now in a position to use these concepts of net benefit as a guide to the selection of the optimum plan for any project.

The Optimum Plan for a New Highway

When proposals for a completely new highway are considered many factors have to be taken into account in deciding whether a new road is worth building at all, and if so what type it is to be and what route it is to follow. The route will largely be governed by topographical considerations, though there might well be more than one practicable route. Various types of highway will be possible on any route; two-lane, three-lane, four-lane, divided, for example, each with different standards of surface, gradient, alignment, and so on. As a result of experience the planning authority will be able to select a certain number of plans as worth consideration, and discard others as clearly unsuitable. Among the latter would be such anomalies as a four-lane divided highway with low-type surface, since if the volume of traffic justifies the one it must rule out the other. The problem therefore resolves itself to one of choosing among a certain number of practicable plans.

Each plan can be analysed in terms of the cost curves and demand pattern to arrive at the anticipated stream of net benefits which would result from its adoption over its anticipated life. Once these are calculated compound plans must be considered. Thus, if traffic volumes are expected to grow in future years, a two-lane highway might yield higher net benefit in the early years and a four-lane highway in later years. Consideration might then be given to immediate acquisition of right of way for a four-lane road but with initial construction of only two lanes, the second two lanes to be built when traffic volume grows to the point where net benefit from the four-lane plan is greater than that from the two-lane. Similarly, consideration might be given to a low-type surface initially to be replaced later by a high-type surface. Such compound plans can be analysed to determine net benefit without great difficulty. If the second stage would be implemented after ten years of an anticipated twenty-five-year-life for the project, the initial costs of acquiring right of way, development, and base construction can be amortized over twenty-five years, and the costs of the second stage of the project over the last fifteen years. Two sets of cost curves then emerge, one for the first ten years and one for the last fifteen years. These are related to the demand patterns in the appropriate years to determine the potential stream of net benefits.

When the net benefits are calculated for each plan, simple or compound, the streams can be discounted to current values. The plan with the greatest net benefit is the optimum solution. It might well be worth making the selection in two stages in some cases, the first stage taking the broad plan for each type of highway, as a result of which the field of selection would be narrowed, and the second stage considering the remaining plans in greater detail, reckoning modifications of each broad type as separate plans.

This technique of selection is not as cumbersome as might at first be thought since the demand pattern will be the same for each plan, the cost curves of a given plan are not difficult to determine, and by restating the technique of analysis in mathematical terms the detailed calculations of net benefit can be handled by computers. The plan yielding the greatest net benefit is the optimum because by definition this is benefit over and above all costs. If any plan yields positive net benefit the highway is worth building, that plan yielding the highest net benefit being the most worth while.

The Optimum Plan for a Replacement Highway

The analysis of a proposal for a new highway to replace an existing facility is based on the same technique, using net benefit as the criterion. However, in this case, as we have seen above, net benefit is that over and above the net benefit which would result from continued use of the existing highway. Since the calculation of net benefit on the existing road takes no account of historic costs it might well emerge in some case that the net benefit of a replacement would be negative in the early years. This is a common situation met in practice where realignment is being considered. If no highway already existed the better alignment would be worth having, but since we already have the poor alignment the additional benefits of the better plan might not warrant the costs of reconstruction. The practical problem in such a case when traffic volumes are growing is to decide when the new road becomes worth the additional cost. The solution to this problem uses the same technique as that for compound plans for a new road. Thus, if the net benefit from the new road is less than that from the existing road for the first five years, giving negative net benefit for the replacement project, but greater in later years, consideration should be given to the compound plan of retaining the old highway for five years and then replacing it. In this case it would be well to repeat the analysis before the replacement in the light of later and more accurate estimates of future demand. Consideration must also be given to compound plans which involve temporary improvements in the old road before it is worth completely reconstructing, and plans which involve the reconstruction of some sections of the old road before others.

Thus, in the case of a replacement project the problems and technique of analysis are essentially the same as for a new highway. There will be a number of possible plans, simple and compound, and for each a figure of net benefit can be obtained. The plan yielding the highest net benefit is again the optimum.

Improvement of Existing Highways

While the most important decisions facing highway authorities are concerned with the construction of new highways, the most numerous are those concerning improvement of existing highways. Improvements can be of numerous types and might involve capital expenditures such as increasing the number of lanes, building grade-separated intersections, or changing the type of surface; or they might involve only a decision to spend more in maintaining a road in better condition. The decision whether to improve an existing road, and if so in what way, rests on the same criterion of net benefit that governs the analysis of new construction projects.

The essential characteristic of any improvement is that highway costs are increased in order that vehicle and users' personal costs (which together we shall call operating costs) might be decreased. The demand for travel over a road will normally be unaffected by an improvement, though a change in costs might well affect the volume of traffic. We can therefore regard the demand pattern as remaining the same for any improvement as it is on the existing road. The changes come in the cost curves, and will be of four types for a typical improvement. There will be a new element of fixed costs arising from amortization of the capital expenditure on the improvement. Fixed maintenance costs might change either upward or downward; upward where the improvement takes the form of a decision to maintain a highway in a better condition, and downward where capital improvement reduces the rate of deterioration. The third change is in variable highway costs which will normally be reduced by the improvement. Finally, operating costs will be lower.

The first stage of the analysis is the usual one of developing a number of plans for improvement or combinations of improvements, which are worth considering, so that a choice among the possible plans can be made. The choice is determined by the change in net benefit which would result from the adoption of each plan. If any plan for improvement would result in increased net benefit then the existing road is worth improving, that plan which would result in the greatest increase in net benefit being the optimum form of improvement.

Cost curves are drawn for the existing road and for the road as it would be with each of the improvement plans implemented. Amortization of historic costs on the existing road is disregarded throughout, but amortization of the cost of any improvement is included in the cost curves for that improvement. Similarly, a demand curve is drawn for each representative hour. It is convenient to have the cost and demand curves drawn on a medium which facilitates their being superimposed in various combinations. Tracing paper is quite convenient for this purpose, and a base of squared paper greatly simplifies the measurement of areas.

Diagram 9

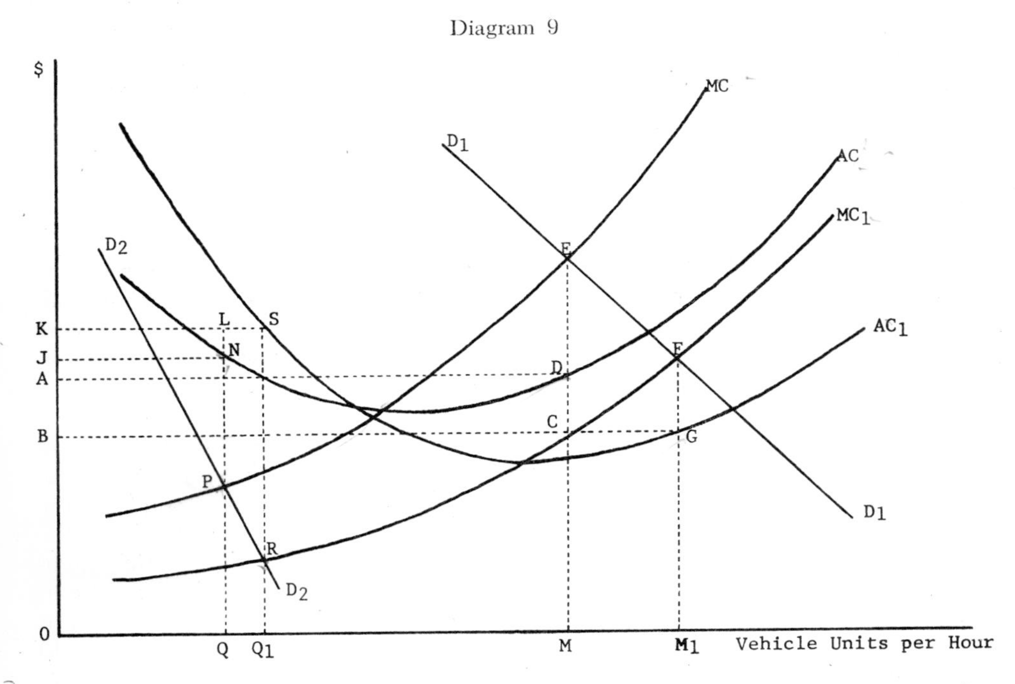

Diagram 9 shows the average and marginal cost curves, AC_1 and MC_1, for one improvement plan superimposed on the corresponding curves, AC and MC, for the existing road. Two demand curves D_1 and D_2, representing two hours, are superimposed on these. In the hour represented by D_1D_1 the optimum volume of traffic on the existing road is OM and average cost OA, while on the improvement the optimum volume is OM_1, at average cost OB. The change in net benefit is composed of two parts. OM traffic, which is common to the existing and improved roads, travels at average cost OB instead of OA, giving a reduction in cost or an increase in net benefit, of $ABCD$. In addition MM_1 traffic now travels, the net benefit from which is $EFGCD$. Thus the increase in net benefit in this hour as a result of the improvement is the area $ADEFGB$. In the hour represented by demand curve D_2D_2, however, there is a reduction in net benefit as a result of the improvement. The existing optimum volume of traffic and level of average cost are OQ and OJ, and those after improvement OQ_1 and OK. OQ traffic now travels at a higher level of cost causing a reduction in net benefit of $KLNJ$, while QQ_1 traffic incurs a negative net benefit of $LSRP$. Thus the reduction of net benefit at this hour as a result of the improvement is the area $KSRPNJ$.

The change in net benefit at each hour is multiplied by the number of hours in the first year for which the demand curve is typical and the results totalled to give the aggregate change in net benefit in the first year as a result of the improvement. Similar calculations are made for the other years of the anticipated life of the improvement and the resulting stream discounted to arrive at a single value for the change in net benefit. Other plans are analysed in the same way. If all plans for improvement result in negative changes in net benefit the existing road is not worth improving. Otherwise, that improvement plan showing the highest positive change in net benefit is the best. If the optimum improvement plan shows a negative change in net benefit in the early years it indicates postponement of the improvement.

New Highway, Replacement, or Improvement?

When faced with the problem of what to do with an inadequate road, a highway authority might be faced with a choice among improvement of the existing road, replacement by a better road, or the building of a new highway while leaving the old in operation. Such a choice can be based on the criterion of net benefit as developed above, by calculating separately the net benefit from each possibility and choosing that with the greatest net benefit.

The maximum net benefit which can be obtained by improvement is the net benefit which would result from adoption of the optimum improvement plan, in excess of the net benefit already yielded by the existing highway. Similarly the maximum net benefit obtainable by replacement is the net benefit from the optimum replacement plan in excess of the net benefit from the existing highway. The third possibility, building a new road while leaving the old in operation, is more complicated. The maximum net benefit obtainable in this way is the sum of the net benefit on the new road plus the net benefit on the old road after construction of the new, and after completion of the optimum improvements on the old road in the light of changed condition resulting from construction of the new road, minus the net benefit yielded by the existing road. The optimum plan of improvement on the old road if a new road is to be built will be different from the optimum plan for improvement if a new road is not built, since traffic volume on the old road will be reduced by diversion to the new. Furthermore, the extent of this diversion will depend on the type of new road built.

Conversely the optimum plan for the new road will depend on the extent of improvement on the old. Thus the alternative plans for the new road and improvement of the old road cannot be considered separately; they are interdependent. Various combinations of plans for the new road and improvement of the old must be considered to show which combination yields the highest net benefit. The technique of measuring net benefit from one such combination under these conditions is illustrated by diagram 10.

Diagram 10

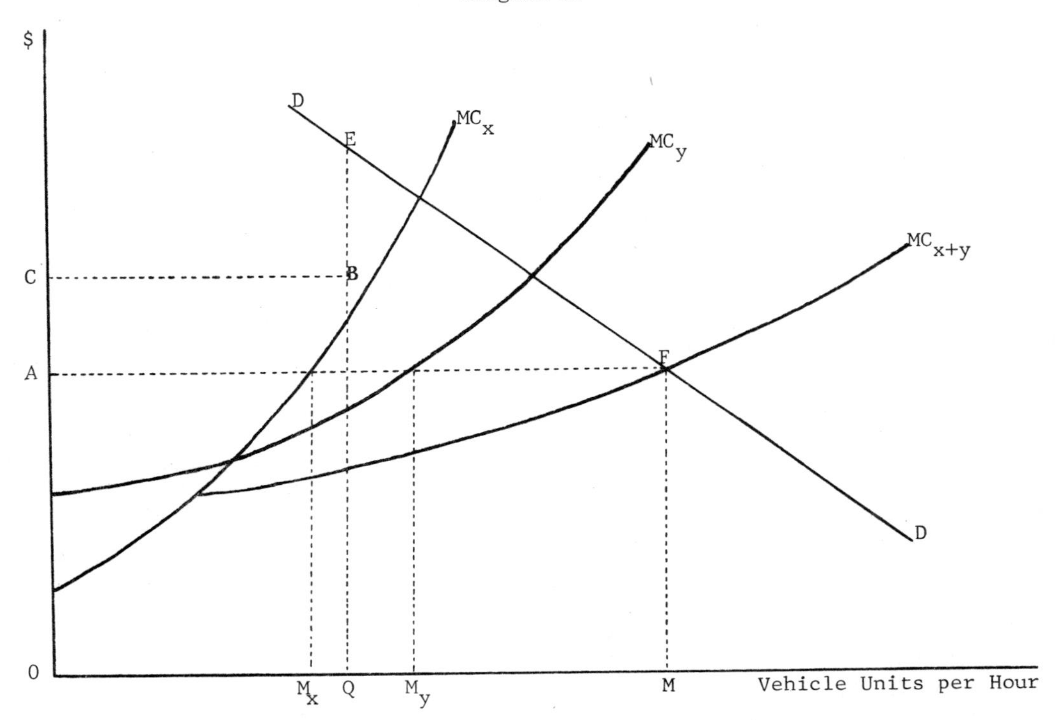

The plans for improvement of the existing road and construction of the new road which are under consideration are each depicted by the normal set of cost curves. The usual practice is observed of including all costs not yet borne but excluding historic costs. Accordingly, all costs are included in the plan for the new road, while the cost curves for the improvement include all costs of improvement but exclude historic costs on the existing road. The optimum distribution of any volume of traffic between the new road and the improvement will be such that the marginal cost on each road is equal. Any other distribution would leave a possibility of reducing total costs by transferring traffic from the road with the higher marginal cost to that with the lower marginal cost. In diagram 10 MC_x is the marginal cost curve for the improvement and MC_y that for the new road. MC_{x+y} is the horizontal addition of these two curves and shows the marginal cost of carrying any volume of traffic if it is distributed in the optimum way between the two roads. *DD* is the demand curve for one representative hour. The optimum volume of traffic on the two roads together is *OM*, at marginal cost *OA*, of which OM_x travels on the improved road and OM_y on the new road. The cost curves of the existing road are not shown, but it is assumed that the marginal cost curve of the existing road cuts the demand curve at *E*, giving optimum traffic volume *OQ* at which average cost is *QB* or *OC*. Thus net benefit from the existing road is the area under the demand curve, above *CB* and to the left of *BE*. Total benefit with the new road and the improvement is the area under the demand curve and to the left of *FM*. Thus benefit on the new road and improvement in excess of net benefit on the existing road is the area *CBEFMO*. To arrive at net benefit we must deduct from this the costs on the new road and improvement, which are the rectangular areas under the average cost curves for the two roads at traffic volumes OM_y and OM_x respectively. The average cost curves are omitted from the diagram to avoid confusion.

In this way net benefit in one hour from the combination of new road and improvement is found. Net benefits at other hours and in other years are calculated similarly and the resulting stream discounted in the usual way to arrive at a single figure of net benefit. Other combinations of new road plan and improvement plan can be examined in the same way, and that combination yielding the highest net benefit is the optimum. Having determined the optimum plan for improvement alone, the optimum plan for reconstruction, and the optimum plan for a combination of new road and improvement of the existing road, the choice among the three possibilities is in favour of that which yields the highest net benefit.

So far we have discussed only some of the planning problems which might face a highway authority. In practice the number of different problems which can arise is limitless, but all choices can be made on the basis of net benefit. Some major problems might well prove complicated, as, for example, in the case of a route between two towns which is considered simultaneously with plans for expressways through or bypasses round the towns. Such projects must be considered simultaneously because the demand for travel on each part is dependent on costs on other parts. However, these relationships, once known or estimated, can all be handled by analysis based on the criterion of net benefit, though the calculations could be simplified by mathematical techniques. When the number of possible combinations is large the detailed calculation can be delegated to computers. Electronic computers are already in use in West Germany[3] for highway planning, and are eminently suitable for the type of calculation involved in this analysis. The costs of undertaking calculations for such a major and complex decision might be considered high, but they would actually be small in relation to the total cost of such a programme and well worth while in the interests of making correct decisions. Adoption of such methods would also avoid the amount of time and effort normally devoted to the political haggling which invariably plagues the highway authority when major decisions have to be made.

Net Benefit with Sub-Optimum Traffic Volumes

So far we have assumed in the calculation of net benefit that optimum traffic volumes are achieved. While it is the function of the pricing system to encourage the movement of optimum traffic volumes, no pricing system can perfectly fulfil this function. The volume of traffic which will in fact flow on a road might therefore be different from the optimum volume. Determination of optimum volumes is worth while since it provides the data on which the pricing system is based, but once that pricing system is determined roads should be planned in accordance with the net benefit which will actually be realized rather than the net benefit which would result from an unattainable optimum traffic flow.

The concept and definition of net benefit remain the same. It is measured for any volume of traffic by the area under the demand curve and above average cost. The only modification necessary to the above analysis to take account of actual rather than ideal conditions is therefore to measure net benefit for the actual volume of traffic which will flow rather than the optimum volume. The actual volume of traffic is determined by the cost of travel to the user in relation to demand price. Cost to the user is composed of vehicle operating costs, users' personal costs, and taxes. The taxation or pricing system recommended in later chapters is in three parts, the registration fee for the vehicle, taxes on components of vehicle operating cost such as the gasoline tax, and tolls. The registration fee does not vary with the number of journeys made by a vehicle and is not therefore a component of the marginal cost of a trip to the user. It is a cost of motoring, however, and its level will affect the volume of traffic. We take account of it by redrawing the demand curve on the assumption that registration fees have been paid. Thus while the demand curve which has been used hitherto shows how many vehicle units are prepared to travel at a series of costs per trip, the demand curve D^1D^1 in diagram 11 shows how many are prepared to travel at a series of costs per trip after the registration fee on each vehicle has been paid. The average operating cost curve (*AOC*) in diagram 11 shows the average cost to the user of one trip for a series of traffic volumes. It is composed of four parts: vehicle operating costs; taxes based on vehicle operating costs such as the gasoline tax, which is reckoned by measuring the cost of fuel at the price including tax; users' personal costs of time, inconvenience, and risk; and the toll per trip in the case of roads where it is intended to levy tolls. If, as is common, users contract out of some of the risks by carrying insurance, the premium on which does not vary with the number of trips, then the component of risk covered might more accurately be excluded from users' personal costs and treated in the same way as the registration fee when drawing the demand curve. To be precise, the costs of vehicle operation which do not vary with mileage might also be treated this way.

The two curves of diagram 11 are now analogous with the normal supply and demand curves of a competitive market, and the equilibrium traffic volume is shown by their intersection. The demand curve will, of course, be different at different times, showing different equilibrium volumes of traffic.

The sole purpose of diagram 11 is to determine the volume of traffic which will actually flow on a given road at any time. Once these volumes are known, net benefit is calculated from the demand and cost curves of previous diagrams, and the anticipated stream of net benefits over the anticipated life of the project discounted to a single figure in the usual way. These figures of net benefit on each possible plan, based on the realistic assumption of actual rather than op-

Diagram 11

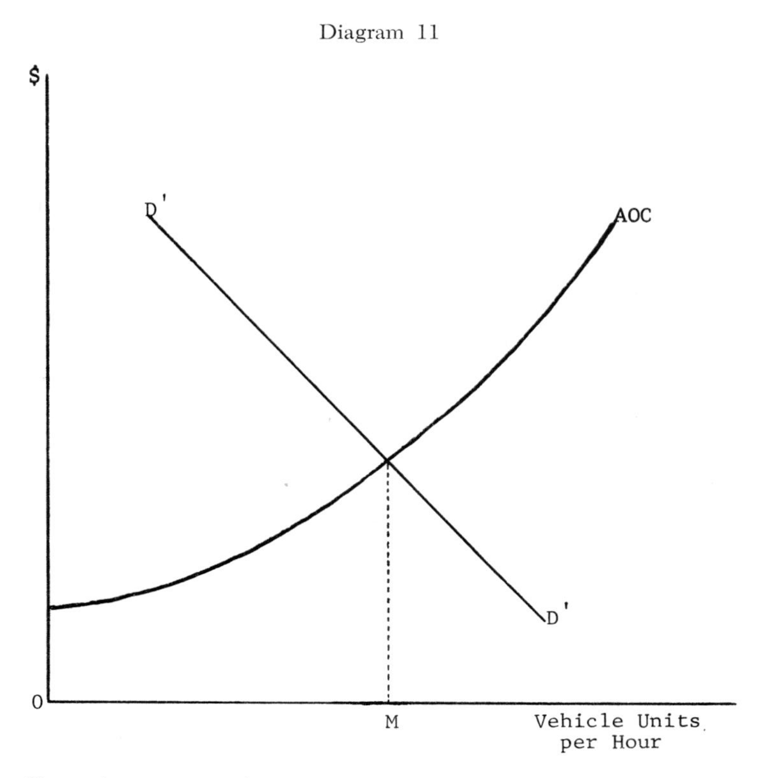

timum traffic volumes, are then used as the criterion for choosing among possible plans.

Priority among Capital Expenditures

When the optimum plans for all projects have been determined, the total amount of capital expenditure involved might be so high that it is not possible to begin all projects immediately. The worth-while projects must then be arranged in order of priority. There are two reasons why some projects have to be delayed, the first being commonly recognized but theoretically of little significance, the second not commonly recognized but of considerable importance.

The first is that the amount of money available for capital expenditures is limited, either because the highway authority must rely on current tax revenues or inadequate allocations from budgets, or because the extent of highway indebtedness or the rate of interest which may be paid on loan funds is limited by law. The use of loan funds for highway construction is discussed in a later chapter where the conclusion is reached that there is no valid objection to loan financing as such. The interest cost of capital expenditures was included in the analysis of costs in chapter II, and if funds are not available on the money market at the rate used in the analysis, then this rate is too low. It is particularly important that a realistic rate of interest be used in the analysis of costs, since the use of the wrong rate can lead to wrong decisions. The lower the rate of interest, the more it is worth spending on long-term investment in highways and the more capital intensive will be the optimum plan for any project. If calculations are based on too high a rate, highways of too low a standard will be chosen and built; in other words, there will be underinvestment in highways. If too low a rate is used, highways of too high a standard will be chosen and funds will not be available for their construction. Delays and the necessity to give some projects low priority will mean overinvestment in some projects and no investment in others. The correct rate is the rate at which the necessary funds can actually be borrowed. However, highway construction requires long-term capital investment, and interest rates are not stable. The plan chosen should be the best from a long-term standpoint and if interest rates are temporarily high it might be better to postpone building a highway of a standard which is optimum from the long-term standpoint than to build a lower-type highway immediately. The converse does not apply, however, and advantage of unusually low interest rates can be taken at any time. Thus, the rate used in the cost analysis should be either the average long-term rate with short period fluctuations removed, or the current rate, whichever is lower. If the cost analysis is based on a realistic rate of interest there will be no shortage of funds available for borrowing, and since there is no objection to the use of loan funds there will be no necessity to assign priorities because of a shortage of capital.

The second reason why immediate implementation of all projects might be impossible or undesirable is that the capacity of the construction industry is limited. This problem is particularly acute where a long period of low expenditures on highways has led to a construction industry with low capacity and at the same time to a large backlog of worth-while projects. The construction industry cannot expand greatly overnight, and there is therefore a maximum rate at which money can be spent on highways determined by the availability of men, equipment, and materials. Where a large backlog of highway construction exists it might not be desirable to encourage the construction industry to expand at its maximum rate, since the industry would then have high capacity at the time when all projects are completed and the rate of construction falls. The result would be a boom followed by a severe slump in the industry. This instability would tend to be recurrent since a high rate of construction over a short period now would mean that many projects would fall due for replacement at about the same time in the future causing another boom then. Thus it might well be both necessary and desirable to spread a construction programme over a few years so as to allow an orderly expansion of the industry to a peak followed by an orderly contraction to a stable level of capacity.

Whether the reason is legal restriction on the rate of expenditure or limitation of the capacity of the construction industry, highway authorities do have to assign priority among worth-while projects. We must therefore establish criteria for determining such priorities. The period for which priorities must be assigned depends on the relationship between the amount which it is worth spending and the amount which can be spent in future years. The planning analysis yields data on how much it is worth spending in the first year, the first two years combined, the first three years combined, and so on, and this amount will grow with time as further projects become worth while which do not warrant immediate construction. Similarly, the limitations on the rate of construction or the availability of capital determine the amount which can be spent in the first year, the first two years combined, and so on. If the amount which can be spent in the immediate future is less than the amount which it is worth spending, priorities must be assigned to cover the duration of the inadequacy. Thus, in the simple example which follows, it is assumed that it is possible to build all worth-while projects in a period of three years, but that the programme cannot be completed in less than three years.

Consideration must also be given to the area over which the programme extends. If the limitation is a financial one the area is that of the jurisdiction of the highway authority, but where the limitation is one of capacity of the construction industry, and the area of jurisdiction of the highway authority is large, separate regional programmes and priorities might be necessitated by immobility of the construction industry. Thus in the extreme case of a federal authority in the United States the limitation of capacity of the industry might be more severe in some states than in others; Californian contractors, men, equipment, and materials cannot be put to work in New England. Different programmes would therefore be necessary for different regions, each based on the amount of worth-while construction in relation to capacity of the industry in the region in question. The size of each region would depend on the mobility of the industry.

The data on which priorities are based can be conveniently arranged as in diagram 12. It is assumed that each project takes one year to complete, and all worth-while projects can be completed in three years. The problem is to decide which projects are undertaken in each of the three years. Those projects undertaken in the first year will be completed two years before those undertaken in the last year. Since all projects will be completed at the end of the third year benefits from then on are not relevant to the determination of priorities. These must be based on net benefit in relation to capital cost over the two-year period when some projects will be completed but not others. The definition of capital cost depends on the factor which precludes immediate construction of all projects. Where money is the limiting factor this term will include all expenditures by the highway authority, but where the limitation is capacity of the construction industry it will include only those expenditures involving use of this capacity. Development and construction costs would be included in the latter definition but not the cost of right of way. Each project to be undertaken in the three-year period is represented by a column in diagram 12. The width of the column represents the capital cost, while the height represents net benefit in the two-year period per dollar of capital cost, this height being in two parts representing each of the two years. Compound projects, such as building the first two lanes of a four-lane road before the other two, are split into two separate columns, the first representing net benefit from two lanes in relation to the capital cost of the first stage, and the second column the additional net benefit of the second two lanes in relation to the additional cost of the second stage. Roads which are not worth building immediately, but will become worth while in the course of the three-year programme, are represented by columns showing net benefit only in the years after their construction is worth while. The amount which can be spent in each of the three years is represented by a horizontal distance, and the problem is to decide which plans

Diagram 12

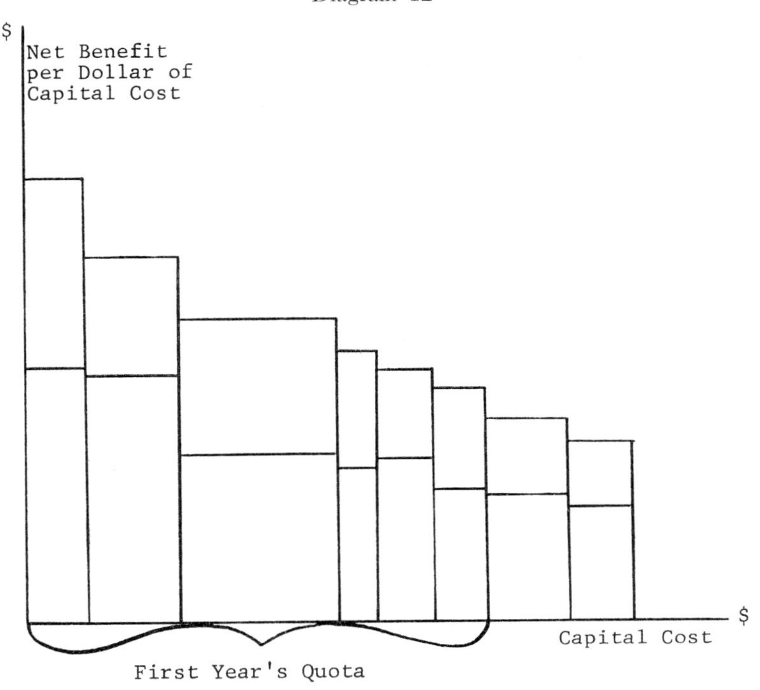

with a combined width equal to this distance should be in each year's quota. The object is to arrange the plans in such a way that the combined net benefit of all plans in the years in which they will be in operation is maximized. This task is simply accomplished by the use of punched cards techniques, but without such devices the process is rather complicated.

The first stage is to assign projects provisionally to the first year's quota in descending order of magnitude of total return of net benefit per dollar of cost for the full two year period, until this quota is full. Of the remaining projects, those with the highest return in the second year are assigned provisionally to the second year's quota. All remaining projects are provisionally in the third year's quota. All projects are then subject to further moves in accordance with the following rules until no further moves are possible. Any project in the second year's quota can replace a project in the first year's quota if the former's return in the first year is greater than the latter's. Any project in the third year's quota can replace a project in the second year's quota if the former's return in the second year is greater. Any project in the third year's quota can replace a project in the first year's if the former's return in the two years combined is greater.

While comparatively simple in this example where there are only three quota periods, the extent of rearrangement can become considerable where more periods are involved, though still quite manageable with punched card techniques. No single criterion for an immediate and final ordering exists. The initial criterion used above for selection of the first year's quota might result in a project being placed in the first year's quota by reason of a high total return over the full period, although very little of this return is in the first year. Some other project with a lower total return, which might be assigned provisionally to the second year's quota, might have a higher return in the first year. Clearly the order of these two projects should be reversed. At the same time we cannot select the first year's quota entirely on the basis of return in the first year, since a project might then be included in the first year's quota although its return in subsequent years is so low that its total return over the full period is less than a project which might, by that criterion, be placed in the last year's quota. Thus in the following very simple case where there are only three projects, all of equal cost, one to be constructed in each of the three consecutive years, either criterion would result in one necessary rearrangement:

Project		*A*	*B*	*C*
Net benefit as percentage of capital cost:	first year	6	5	4
	second year	6	8	10
	TOTAL	12	13	14

Using the criteria suggested above the provisional order would be *CBA*, but since *B* has a higher return in the first year than *C* the final order would be *BCA*. If the first year's project is selected on return in the first year alone the provisional order would be *ACB*, but since *B* has a greater return in the full period than *A* the final order would again be *BCA*.

While there is no single criterion for immediate ordering there is only one necessary condition for a final ordering. No two projects should be so ordered that the return on the latter is greater than that on the former during the total period between their completion dates. The stages of a compound project must, of course, be in the correct order.

We have assumed that all projects take one year to complete. The possibility of a project requiring more than one year will normally arise only if the capital cost of that project is more than one year's quota, or if there is some immobility in the construction industry. In the latter case regional priority orderings should be used. Where a project which will take two years to complete must be fitted into a priority system the initial assumption can be made that all the expenditure will be in the second year. Once priorities are assigned on this basis the choice is faced of moving half the expenditure forward one year at the expense of delaying projects in the earlier year's quota, or moving half of the expenditure back one year at the expense of losing one year's return on the large project, but with the gain of one year's return on projects which can simultaneously be moved one year forward. Whichever of these possibilities involves the smaller sacrifice of net benefit should be adopted.

A number of other tacit assumptions have been made, the most serious of which is that the level of net benefit on each project is independent of the construction of others. Where a large construction programme is under consideration the net benefits from the various projects might well be mutually dependent. Thus an expressway between two towns might be planned simultaneously with the construction of throughways or bypasses at the ends. Delay in any part of this project would affect the level of benefit on other parts. The number of possibilities of relationships of this type existing between projects is virtually unlimited, and it would be impossible to discuss each in detail. Once the relationships are known, however, the criterion of maximum net benefit can still be applied. Of all possible priority orderings the optimum is that which yields the greatest net benefit. The difficulty of determining which ordering is the optimum is simply one of lengthy calculations, and these become quite practicable with the availability of computers and punched card techniques.

One great advantage of the criterion of net benefit which has been used throughout the analysis of planning and priority ordering in this chapter is that it is applicable not only to highways, but to all forms of public works. The techniques of analysis might differ with different cases, but the same criterion can be used. This means not only that highways can be planned rationally rather than as instruments of political policy, and that similar methods can be used for other public works, but that the relative claims of various forms of public expenditures can be assessed by reference to a common criterion, and priorities established not only for each form of expenditure but between forms. While the administrative costs of employing such techniques on a scale which would embrace all public works would be considerable, the end result would be so valuable to society that these costs would be small by comparison.

The Effects of Different Conditions on the Optimum

Having analysed the way in which costs and demand theoretically interact to determine the optimum plan, it is worth pausing at this stage to relate the theory to some examples of the influence which different conditions might in practice have on the optimum.

Topography will naturally play an important part in determining costs of construction. Other factors being the same, the costs of construction will increase the more mountainous the country, and up to a point they will increase at an increasing rate. Where a road crosses contour lines slight gradients might make little difference to operating costs, but with steeper gradients more work on grading, cuttings, and embankments will be worth while, because after a point vehicle costs increase with gradients more than the construction costs necessary to avoid them. This will considerably increase costs of construction so that only two lanes might be worth while whereas four lanes would be the optimum in level country. Where the road follows the contour along the hill-side, grading costs will increase proportionally with the gradient if soil conditions permit vertical sides to cuttings, and more than proportionally if, as is usual, sloping sides are needed. The more gentle the maximum slope of cutting and embankment sides the more costs will increase with gradient. In any case grading costs will increase with the square of the width of the road surface. Other factors being equal, there will therefore be a greater tendency to dual carriageways on lateral slopes since these can be at different levels. Grading costs will therefore be twice those for a single carriageway rather than four times as would be the case of a single carriageway of twice the width. These relationships follow from simple geometry on the assumption that grading costs are proportional to the amount of soil moved. Roads will therefore be narrower, fewer in number and of lower standards of alignment both vertical and horizontal, the more rugged the country, for a given demand pattern. However, where gradients have different effects on the speeds of different types of vehicles, and poor vertical alignment restricts sight distance for overtaking, congestion costs might be so high that additional lanes are worth while on gradients but not on level sections of road. The lower the cost of improvements, and the greater the benefit which they will yield, the more they will be worth while. There is no standard of width, gradient, or alignment for a given demand pattern; the optimum plan depends on costs as well as demand.

Other natural conditions will similarly affect both construction and operating costs. A wet climate may increase operating costs on low-type surfaces so much that improvements are worth while for a volume of traffic which would not make them worth while in a drier climate. Similarly, frost may cause such high maintenance costs on high-type surfaces that lower-type surfaces might be the optimum for traffic that would justify a high-type surface in a milder climate. Various soil and subsoil conditions will affect construction, maintenance and operating costs to differing extents, and what is the optimum for one region might not be the optimum under otherwise similar conditions in another region.

Economic conditions in an area will be a very important determinant of the optimum type of highway. Right of way will generally be more expensive the more developed the area and narrower roads will therefore be optimum for given volumes of traffic. This will usually be offset in practice by the greater demand in developed areas. A striking comparison which illustrates the importance of land values in total costs is that the estimated cost of London's Inner Ring road is ten times as high per mile as that of the proposed tunnel under Mont Blanc.[4] If demand for use of both roads were the same it might therefore be optimum to have, say, a two-lane ring road in London and twin two-lane tunnels under Mont Blanc.

Underdeveloped countries will usually have comparatively low labour costs, but a high rate of interest because of the numerous competing demands for available capital. Because wages are low, time and inconvenience will be less valuable and roads which require less capital expenditure, but have lower design speeds and poorer surfaces, might therefore be optimum for a volume of traffic which would make higher standards worth while elsewhere. Demand conditions will also be different in underdeveloped countries. A greater mileage of low-type highway might yield greater benefits than a lesser mileage of high-type highway at the same construction costs, while the reverse might be true in Western Europe. This is largely because anticipated traffic volumes will be different. Where a country has effectively no highways, the first one will have an anticipated demand composed entirely of generated traffic, since there will be no existing traffic; there can therefore be no secular growth, and there are no other roads from which traffic can be diverted. Generated traffic is the most difficult to forecast and the demand curve will therefore be subject to large margins of error. A low-type highway which can be improved later if demand warrants it might therefore be the best plan. This will be reflected in the analysis by a very short anticipated life because of forecasting difficulties. Average fixed cost will therefore be high and the optimum plan will be one with low capital expenditure.

Population density will influence the optimum in three ways. The more sparse the population the lower are the community costs, because there are fewer people to benefit or suffer indirectly from highway development, and less trade to be diverted. The denser the population the greater is the demand, but also the higher are the right-of-way costs. Invariably demand prevails as the more important determinant, and the greater the population in any region, the greater the proportion of total area devoted to highways and other means of transport. This results from the fact that with increasing population density the costs resulting from congestion increase faster than the cost of right of way.

The economic prospects of a region also affect the type of highway that will be optimum. If prospects are poor, anticipated life is short, annual fixed costs are high, and a low-type highway is built. Conversely, if prospects are good, a higher-type highway will be built. If conditions turn out to be better than anticipated, a region will find itself with inadequate highways already largely paid for. If conditions turn out to be poorer than anticipated, there will be good highways but a heavy debt burden on historic costs.

The demand for highway use will be influenced by geography, economic development, the density and distribution of population, location of resources and markets, and the general level of propersity. Demand can vary in many ways; not only is the demand for highway services greater in the United States than in Africa but the traffic composition is different. The formula for reducing traffic to standard vehicle units will have to be different for different conditions, because of the different relationships between costs. This partly explains the different approach to highway design in different countries and at different times. The Romans usually went over hills in a straight line, while the English went round them. The Americans move the hills. Roman roads were built mainly for foot traffic which was affected more by distance and less by gradient than vehicular traffic. Further, Roman wheeled carriages, drawn by animal or human power, had fixed axles making straight alignment important. English roads were built mainly for stage-coach traffic where curvature was relatively unimportant, and gradients much more of an obstacle than increased distance. Expressways in the United States are built for high-speed motor traffic where gradients, curvature, and sight distance are important. These factors combined with the reduced construction costs brought about by mechanization make it worth going to much greater effort to secure both a good horizontal and a good vertical alignment.

The elasticity of demand is important in estimating future traffic. Generated traffic depends on the elasticity of demand for travel by any means, while diverted traffic depends on the availability and costs of other forms of travel. The greater the variety of forms of travel available, the greater the elasticity of demand for highway use, and the greater will be the effect of a given change in operating cost on the volume of traffic.

Fluctuations in demand are important in determining what will be the optimum type of highway for a given annual average daily volume of traffic. Generally speaking, congestion costs increase at an increasing rate, so that irregularities round the average lead to greater increases in costs at times of heavy traffic density than decreases at times of light traffic. The average cost per trip is therefore higher the greater the fluctuation round the average. A given improvement which reduces costs caused by congestion will therefore yield greater benefits the greater the fluctuation in traffic density. Thus for a given average traffic volume, a higher-type highway will usually be more worth while the more irregular the traffic density.

These examples could be extended indefinitely, but these few are adequate to demonstrate the importance of planning the road for the conditions of the area. No single factor determines what type of highway is the best, it is determined by the interaction of many forces. We have shown how these forces are related and how the optimum plan can be determined for any given set of conditions. The next chapter examines how we can measure the importance of the various factors in a given case.

CHAPTER FIVE

THE EVALUATION OF DATA

IN PREVIOUS CHAPTERS we have assumed throughout a perfect knowledge of all the data required in the analysis. Now we must examine the methods by which these data can be collected and converted to the units of money and traffic volume on which the analysis was based. We begin with the costs, taking them in the order in which they were examined in chapter II.

FIXED HIGHWAY COSTS

Right of Way

The choice of location of a road is one of the most important decisions facing the planner. Right of way will vary greatly in value between alternative routes, and it is vital when comparing possible locations from an economic standpoint to put a true value on land used. The real cost of using any piece of land is the loss incurred by not having that land available for other uses. This can only be measured by the current market value. It is often advocated that land be purchased by highway authorities in advance of requirements to avoid increases in its market value which follow announcements of development of the area by highway construction.[1] This may be sound practice from a financial standpoint, but it should not confuse the valuation of the land at the time of construction. The value of a strip of land for other than highway uses today is not directly affected by whether it is to be purchased today or was purchased five years ago. It is the value today which is important.

We must, however, distinguish between the value of land in an area if a highway is to be built, and the value if it is not. If a new highway is built across level virgin country the value of adjacent land will increase, and if the highway could be built either to one side or other of its proposed route one cannot distinguish the adjacent land from that actually used. The relevant value is, therefore, that of the land if no highway is to be built, since that is the value in other uses. If we take the value of adjacent land, which value takes account of expected development, we are really assuming that the opportunity cost of the land used is its value with a highway adjacent, which is much greater than the value of the land if no highway is built. Thus, if right of way was acquired well in advance of requirements, the cost of using it for a highway is neither what it was worth when bought, nor what similar adjacent land is worth now; it is what the land would be worth if sold now for other purposes and no highway were built. This is best determined from data of past land sales in the area, adjusted for any rise in value over time caused by anything other than potential highway development. Any increase in the value of adjacent land is a negative community cost to be discussed later.

Where land that is already developed has to be acquired, the problem is more complicated. This arises particularly in cases of widening where a strip of land might be needed from adjacent sites. One method of valuing this which is widely used is to take the value today of the whole site and deduct the value of the remainder after construction. This really combines two costs, however: the cost of the land actually used plus the change in value of the remainder as a result of the construction. As we shall see below, including the value of betterment to adjacent land involves serious problems of double counting. At this stage, therefore, we wish to exclude betterment from the value placed on right of way. This can be done if the value of the remainder of the site can be assessed before the road improvement is announced, so as to exclude speculative rises in value caused by the proposed construction. An alternative method which achieves this is to take the market value of the land actually used if it were sold for non-highway purposes. This is also open to objection, however, since it would involve loss of access to the remainder of the site. What we are really trying to determine is what would be the value of the strip of land if the present highway were moved, without being improved in any way, so as to give the same frontage to the remainder of the site. In the case of farm land this will normally be equal to the value of a similar strip of land on the other side of the site, i.e., that farthest from the highway, if that were sold for non-highway purposes. This can be assessed from past sales of similar land.

The cost of right of way includes not only the value of the site but also any buildings or other structures on the land. Thus, if we take a strip of land twenty feet wide from the frontage of a site and all buildings apertaining to that site are on those twenty feet, the true value for our purposes will be the value of the land and the buildings on it. Where buildings extend more than twenty feet it will often be impossible to demolish part of the buildings without demolishing them all, and the full value of the buildings must therefore be included. Since it is impossible to divorce the value of a building from that of the site, this will equal the total value of site and buildings less the remaining value of the remaining site. Where it is possible to leave some part of the building, its remaining value must be deducted together with the value of the remaining site. All these components can be combined for evaluation purposes by a simple procedure. The value of the land used will be the value of the total site and buildings less the value of the remaining site and buildings or parts of buildings (where these have any value) with access to a highway of comparable standard to that before the improvement.

In built-up areas buildings are the major part of right-of-way costs. It is interesting to note, therefore, that the cost of a widening which involves demolition of buildings will be very high, while a further widening at the same time would have much lower costs. Thus, to take a five-foot strip might be possible without incurring damage to buildings, the next five feet might necessitate heavy damage to the buildings, and a further five feet might involve their complete demolition. The cost of still a further five feet might then be very low. This is a very important consideration where it is possible to widen a highway on either side or partly on each. If all buildings along the route are of this type and one extra ten foot lane is needed, it will be cheaper to take five feet from each side. If two extra ten-foot lanes are needed, however, it will be cheaper to take it all from one side. In practice all buildings will not be of the same type and size, and a determination of the best plan will then make it vitally important to assess accurately the exact cost of right of way needed by each possibility.

A precise valuation of right-of-way costs is also vital where we are faced with the alternatives of widening roads to build an expressway, or building a bypass on cheaper right of way but at higher construction cost. The optimum plan will of course depend on all the relevant costs and the traffic pattern, but this can only be determined if each is assessed accurately. Two examples of this illustrate the point.

Three miles of Woodward Avenue, Detroit, were enlarged at a cost of $11,000,000 of which $9,800,000 was for land and property damage. Approximately eleven miles of controlled access highway on less expensive right of way could have been constructed with these funds, with far greater improvement in functional service. The estimated cost of land for widening the Albany Post Road in Westchester County, New York, from 66 to 166 feet would have been over $792,000 per mile, while land in the same county for the Saw Mill River Parkway of controlled access design on an entirely undeveloped new location and averaging 500 feet in width cost only $138,000 per mile.[2]

These figures are of compensation paid, which may not be the same as the true opportunity cost of right of way, and Mr. Levin's opinion in the Detroit case might or might not be borne out by a full study. But they do point to the importance in planning major improvements both of a complete and accurate evaluation of right-of-way cost, and of basing this evaluation on sound economic principles.

Such evaluation of the cost of land used is as vital in more detailed questions of alignment as in the choice of an over-all route. An urban expressway can be built to varying standards of alignment, and in general the better the alignment the higher the right-of-way costs and the lower the vehicle operating costs. Similarly interchanges of various types require different combinations of right-of-way requirements, construction costs, and vehicle operating costs. The optimum can only be determined by a sound and accurate evaluation and comparison of these.

By valuing right of way at its market value for other uses if no highway improvement were undertaken, account is taken of the impersonal attributes of the land such as its situation and business connections. Where there is nothing unique about the site in question, market value will not be difficult to determine, and the displaced occupier will not find it difficult to find another site equally suited to his purposes. It will still be true, however, that the value of the land to its present occupier will be higher than its salable value; otherwise he would already have sold it. Where there is a ready free market in sites of this type the difference will be the cost of disturbance, including removal and the losses incurred in building up new business connections. Where a site or building has such unique characteristics that the occupier cannot find a similar replacement, disturbance costs will be higher. These will be difficult to determine as there is then no market criterion. The value of an historic family home is more to the occupier than to anyone else. Similarly, a site with rare natural resources or an exceptionally advantageous location for a particular purpose may have no substitute. Such cases are very rare, and valuation of losses the function of civil courts. If compensation is offered at assessed salable value with the right of appeal to the courts, experience of past settlements would enable the highway authority to make some arbitrary valuation for planning purposes in any new case. Valuation for other purposes, such as the total loss of an historic building by fire, might provide guidance. Actual compensation would, of course, be a matter for the courts in each individual case.

Thus we can conclude that the true value of right of way used for highway purposes is the total loss by reason of the land ceasing to be available for other uses. Where a whole site is needed this is measured by the market value of a similar site and buildings plus all costs direct and indirect incurred in moving the occupier. Where only part of a site is needed the cost depends on whether the remainder is still suitable for its original occupation. If it is, the value will be the difference between the value of the whole site before the highway improvement, and the value of a site similar to the remainder with access similar to that before the improvement. If it is not, the value will be the market value of a similar site plus disturbance costs, less the value of a site similar to the remainder on a highway similar to that before improvement. Values in all cases are assessed by surveying practices in relation to data of recent land sales in the area. This valuation for purposes of analysis is quite distinct from the assessment of compensation, which is a financial problem to be discussed in a later chapter.

Development Costs,[3] Construction Costs, Maintenance, and Administration

These costs were described in chapter II. Their determination, in so far as they are fixed, will be a straightforward procedure on the same lines as that used by the construction industry for costing tenders. Administrative costs will not be difficult to estimate, and property taxation will be at the normal rate in the area on the property acquired.

Variable Highway Costs

The functional relationships among the volume of traffic and the costs of depreciation, maintenance, and traffic control systems are largely technical matters which are at present receiving much attention. A series of large-scale experiments in the United States by the American Association of State Highway Officials and the State Highway Departments has accumulated much useful information. More research is needed, but the problems of determining from a technical standpoint what are the functional relationships between highway costs and traffic volumes do not appear insoluble. Estimates of the variable highway cost curves could be made in the light of present knowledge by normal costing techniques, and the degree of accuracy possible will increase in the future with the accumulation of further knowledge and experience.

Vehicle Costs

The costs of vehicle operation vary in practice with many factors, which for our purposes fall into three categories: the composition of traffic; highway conditions; and the degree of congestion. These have been represented on one diagram by reducing the different types of vehicle in the traffic composition to standard units by a weighting formula, and measuring the number of units on the horizontal axis. This takes account of both the composition and volume of traffic. The vertical axis measures costs, and a different curve is used for each highway. The factors governing each of these deserve closer examination. We shall consider first the effect of different combinations of vehicle types and the way these can be reduced to a standard unit; secondly the effect which different highway conditions will have on the vehicle costs for a given traffic composition; and finally the way in which vehicle costs vary with the volume of traffic.

Traffic Composition

Hitherto we have assumed that a single formula can reduce any given vehicle into a number of standard vehicle units, with standard cost incurring potentialities. In practice the weighting will be a function of many things, especially highway conditions. It is convenient, however, that in those cases where commercial vehicles have very high costs in relation to passenger cars this tends to be the case for all types of cost. Thus on steep gradients operating costs will increase more for commercial vehicles than for passenger cars, and so will congestion costs, highway costs, noise, smell, and the like. A convenient rule of thumb was laid down by the Public Roads Administration,[4] "In relation to highway capacity, one commercial vehicle has approximately the same effect as two passenger cars in level terrain and four passenger cars in rolling terrain. In mountainous terrain the effect of one commercial vehicle may be as great as eight passenger cars." Further research is needed to give weighting factors for different classes of commercial vehicle under different conditions of gradient, alignment, and so on. Probably the most important relationship is that between vehicle type and congestion as reflected in vehicle and users' personal costs. Where the weighting factors for different cost curves would be different, the factor appropriate for vehicle and users' personal costs should therefore govern the formula used for the axis. Adjustments to this can be made for the other curves by reference to the composition of the demand pattern. Although the demand pattern is expressed in standard vehicle units, data will be available of the actual numbers of vehicles in each class. If a 2 : 1 ratio exists between commercial vehicles and passenger cars for congestion purposes, while a 4 : 1 ratio exists in damage to the highway, the average variable highway cost curve can be weighted for the fact that, say, 20 per cent of anticipated demand in standard traffic units is from commercial vehicles. Thus if the passenger car is the standard unit, a flow of 100 vehicles per hour on the horizontal axis will be made up of 80 cars and 10 trucks. The effect on congestion will be that of 100 cars, but the effect on highway costs will be that of 120 cars. Thus the point on the average variable highway cost curve corresponding to 100 vehicles per hour will be the total variable highway cost of 120 vehicles per hour divided by 100. All curves can be similarly adjusted where the discrepancy is significant. In practice it will often be so small as to be negligible, and will be less important the higher the proportion of traffic made up of passenger cars.

The disadvantage of this method of compensating for the different importance of different types of vehicle on different costs by adjusting the curves is that the final set of cost curves is drawn with relation to one demand pattern. This is no disadvantage in so far as the analysis only calls for its use in relation to that demand pattern, but may cause difficulty where the pattern varies greatly over time. Thus if the percentage of commercial to total traffic is higher at night, precise theory would call for a separate set of cost curves reflecting this. There is no theoretical obstacle to having a different set of cost curves for every time of day, but in practice the inaccuracy caused by using an average traffic composition will almost certainly be within the error margins of the data, and further complications of the theory will not normally be worth the effort involved.

Highway Conditions

The most important highway conditions which affect costs for any given type of vehicle under any given degree of congestion are: type of area; type of highway; speed; gradient; surface; and alignment. These must be considered for the four basic classes of vehicle; passenger cars, single-unit trucks, combination trucks, and buses.

The type of area will affect vehicle costs because of its effect on traffic conditions. A distinction is normally drawn between urban and rural operation and the former is in all cases the more expensive. Fuel consumption of a passenger car in New York City is approximately twice the level on rural roads. Data of resale value show that the depreciation rate of vehicles in New York City is 2.8 per cent per month compared with a national average of 2.4 per cent. Maintenance costs are approximately 50 per cent higher per mile in urban than in rural areas. This classification into rural and urban is useful in assessing the less direct costs of operation, such as maintenance and depreciation, which are difficult to allocate between vehicle miles on different roads. In all cases, however, the difference in costs is related only indirectly to the type of area; the direct cause is highway conditions. Thus, in estimating what vehicle operating costs will be on a proposed highway, it is the expected conditions on the highway which are important rather than those in the surrounding district.[5] The type of locality affects highway conditions in so many ways, however, that when estimating costs by analogy with an existing highway of similar type it is important to choose one in a similar area. Even the number of pedestrians crossing the road can have an appreciable effect on vehicle costs.

The type of highway is clearly of great importance to vehicle costs. In this context we exclude such properties as gradient and alignment from the classification of types of highway limiting the term to the number and arrangement of lanes, and the degree of control of access. Two-lane highways have higher capacities and lower vehicle operating costs than single-lane highways. Sight distance, which is largely a function of alignment and gradient, is of great importance on two-lane highways because of its effect on overtaking. Curvature will increase vehicle costs on any road by necessitating either reduced speed or increased tire wear, but on two-lane roads particularly it will have an additional effect on operating costs by restricting sight distance for overtaking.

A three-lane highway has considerably greater capacity than a two-lane highway because overtaking does not necessitate gaps in the opposing traffic stream Capacity and therefore speed and costs are greatly affected by the arrangement of the lanes, and whether clear priority is given on any section to overtaking traffic in one direction. Where this is done the effect is that of half a two-lane and half a four-lane highway. Where it is not, two alternative effects arise with moderate to heavy traffic. Either the centre lane is not used for overtaking to its full extent because neither stream has priority, or both use it and the accident rate rises. Thus the capacity of a three-lane highway without directional priority on the centre lane is little greater at high traffic volumes than a two-lane highway.

Four-lane highways overcome the overtaking problem, but their capacity and speed for a given volume are found in practice to be little higher per lane than on two- or three-lane highways unless there is a median divider, when the difference is significant, probably due to the effect of an increased sense of safety on driver behaviour. Divided highways with more than four lanes have much the same capacity per lane as four-lane divided highways.

Control of access can make great differences in vehicle operating costs, by savings both in time and running costs. The capacity of an urban motorway is approximately 2½ times that of a road of equal width without access control. This is shown by the figures in table I.

TABLE I

PRACTICAL CAPACITY OF FOUR-LANE HIGHWAYS[6]

	Urban *street*	Rural *motorway*	Urban *motorway*
Passenger cars per hour	1,250	2,000	3,000
Percentage urban street capacity	100	160	240

SOURCE: Max Erich Feuchtinger, "Some aspects of the design and planning of urban motorways," British Road Federation, Urban Motorways Conference (London, 1956).

The effects on vehicle costs of the greater capacity of controlled-access highways are illustrated by the results of a series of tests in the Boston area of Massachusetts in 1953 (see table II).

TABLE II

EFFECT OF ROAD TYPE ON SPEED AND FUEL CONSUMPTION

Type of road	Average speed m.p.h.	Average fuel consumption m.p.g.
Congested city streets	8.0	9.1
Intermediate city streets	17.7	15.0
Old two-lane highway	22.6	17.1
New limited access highway	46.6	18.3

Economies in operating costs of similar magnitude have been found wherever comparative studies have been made. Fuel consumption on the Merritt Parkway at 52 m.p.h., is approximately the same as on the Boston Post Road at 25 m.p.h., the economy of steady running offsetting the diseconomy of high speed. Tests in Iowa showed one stop and restart to cause as much tire wear as one mile of normal operation, and use enough petrol to run two blocks. From data collected under laboratory conditions (see table III) one can calculate the cost of stops in terms of fuel, maintenance and time.

TABLE III

EFFECT OF STOPS ON FUEL CONSUMPTION

Speed m.p.h.	No stops		5 stops per mile	
	m.p.g.	Fuel cost/mile	m.p.g.	Fuel cost/mile
25	23.867	1.17¢	15.748	1.78¢
30	22.750	1.23¢	13.600	2.06¢
45	20.000	1.40¢	9.685	2.89¢

SOURCE: Lloyd Aldrich, *The Economy of Freeways* (Los Angeles, 1953).

On comparable lengths of freeway with no stops and highway with five sets of traffic lights per mile, average speeds would be in the order of 45 m.p.h. on the former and 30 m.p.h. on the latter with comparable traffic volumes. The difference in fuel cost per mile is 2.06¢—1.40¢ = 0.66¢ or 0.13¢ per stop. In practice synchronization of traffic lights reduces stops on an urban road to an average of two per mile, but slowing down can equal an extra half stop per mile in fuel cost. Halving the number of stops on the above figures gives a net saving on fuel alone of 0.33¢ per vehicle mile by limited access. The saving in time due to the higher average speed will be 0.67 minutes per vehicle mile.

Brake and clutch wear is entirely due to stop and go travel. A survey of service stations in the Los Angeles area estimated average expenditure of $25 per 25,000 miles on brake lining and adjusting and $35 per 35,000 miles for clutch repairs giving a total of 0.20¢ per mile. Freeway travel could eliminate 90 per cent of this or 0.18¢ per mile. Tire costs in Los Angeles average 0.27¢ per mile and other maintenance on moving parts 0.25¢ per mile. It is estimated that 12.5 per cent of these is due to stop-and-go travel, giving a further saving of 0.06¢ per vehicle mile.

On these conservative estimates, which are some years out of date, we find the total saving by elimination of stop-and-go travel to be 0.33¢ on fuel, 0.18¢ on brake and clutch wear, and 0.06¢ on tires and maintenance, a total of 0.57¢ per vehicle mile plus 0.67 minutes per mile. Valuing time at only $1.00 per vehicle hour gives 1.11¢ for 0.67 minutes and a total saving of 1.68¢ per vehicle mile. Time savings are often the most important economy of motorways, particularly in urban areas. On the Davison Limited Access Highway in Detroit a 1½-mile trip which previously took 20 to 30 minutes now takes 3 to 4 minutes. The 25-mile Willow Run Expressway in Detroit saves 5,000,000 man-hours a year in travelling time. The Arroyo Seco Freeway from Pasadena to Los Angeles saves 25 minutes on a six-mile trip.

Speed of operation will have a considerable effect on vehicle costs. Mention has already been made of the effect of variation in speed in stop-and-go travel, but under given conditions vehicle operating costs will normally be higher at higher average speeds. Minimum vehicle operating costs per mile are generally achieved at the lowest speed at which a vehicle can operate smoothly in top gear, which is normally between 20 and 25 m.p.h. Data are readily available of the costs of operation of representative vehicles at various speeds.

Speed of operation also has an effect on those vehicle costs which do not vary with mileage. These are a function of time and can be expressed in dollars per hour. The higher the speed of travel, the lower are these costs for a given trip, this economy of speed being realized in the fact that at higher speeds the same amount of work can be performed by fewer vehicles. This is of little importance in the case of private passenger cars, since reduced travel time is likely to result only in the vehicle spending more time idle; but it is of considerable importance in the case of commercial vehicles where the time saved will normally be put to productive use. Only the time cost of the vehicle itself is included in vehicle costs; the time of the driver and passengers is part of users' personal costs which are discussed below.

The conclusion drawn from examination of numerous speed studies is that there is a definite relationship between the type of highway, volume of traffic, and speed of operation. Given details of the proposed highway, we can estimate average vehicle speed for any volume of traffic, and so take account of the effect of speed on vehicle costs in expressing these as a function of traffic volume.

Gradients can have a considerable effect on vehicle operating costs in many ways, such as through reduced speed, increased fuel costs and brake wear, and increased congestion caused by heavy vehicles. There are many factors governing the extent of these effects, the most important of which are: rate of grade, length of grade, combination of grades, over-all rise and fall, approach conditions, terminal conditions, power of vehicles, and the necessity to change gear or brake. Attempts to combine these into a simple measure of gradient suffer from varying degrees of inaccuracy. Over-all rise and fall takes no account of the rate or length of individual grades. Average gradient has similar faults unless rates of grade are very uniform. Total lengths of grades in classified groups take account of the rate of gradient but not the length of individual grades. Further research is needed to determine a formula which gives due importance to all the aspects of gradients. Data are available of such items as fuel consumption for representative vehicles on different grades, and where data are not available they can easily be collected. The problem of assessing the direct effect of gradient on vehicle operating costs by formulae taking all aspects of the problem into account is a technical one which should not prove insoluble with further research. Data collected by the Road Research Laboratory reach the conclusion that average speeds are reduced by 1.37 m.p.h. for every degree of average gradient, but figures for particular grades under particular conditions vary widely.

The effect of gradient on congestion is more difficult to assess, and varies with the rate and length of grade, alignment, number of lanes, and volume and composition of traffic, all of which interact. Thus on a two-lane road with perfect alignment and low traffic volume a heavy truck will cause little congestion because passing is easy. Reducing gradient by increasing curvature might reduce sight distance and overtaking possibility, and therefore reduce average speed even though it increases the speed of the vehicle causing the congestion. Where traffic is so heavy that passing would be difficult anyway, a lesser grade with greater curvature might increase average speeds. Sufficient is known about the effect of gradient on costs to make it quite possible to estimate the effect which gradients on the proposed highway will have on vehicle costs at any level of traffic, given the specification of the highway and data of traffic composition.

Surface type and condition will affect vehicle costs in three ways. Poorer surfaces will cause higher tractive resistances and therefore cause higher fuel consumption; increased resistance and rough surfaces will lead to increased mechanical wear and tear and therefore higher maintenance and depreciation costs; and poorer conditions will reduce speeds and increase time costs. A great deal of technical research has been done on this question, and the results of various studies have fairly uniform results. More work remains to be done especially in establishing a standard classification of surface types, and in examining the effect of other variables such as gradient and climate, in conjunction with different surface types. The state of technical knowledge on this problem is sufficient at present to estimate the effects which surface type will have on vehicle costs, but the accuracy of such estimates will doubtless increase with the acquisition of further knowledge and experience. Surface conditions also affect the users' personal costs of time, inconvenience, and risk, which will be discussed below.

Alignment of the highway will affect vehicle costs in three ways; increased curvature will increase tractive resistance; centrifugal force will cause greater tire wear; and slowing down for curves will increase fuel, brake, tire, and time costs. So many variables are involved here that the net effects are difficult to assess. Research has shown the extent of some of the effects under laboratory conditions, but much depends on driver behaviour. If a curve is taken too fast tire wear is excessive, but excessive reduction in speed leads to very high fuel and time costs. The American Association of State Highway Officials has devised a system of added percentages to vehicle costs for curvature, by a formula taking degree of curvature and superelevation into account. This is doubtless the easiest way to include the effect of alignment in vehicle costs, but the exact correction factors needed in a particular case will depend on observed driver behaviour, and may well differ for different types of road in different areas. The magnitude of the effect is estimated by the A.A.S.H.O. for degrees of curvature normally used as up to 15 per cent increase in fuel cost and up to 200 per cent increase in tire cost over conditions on straight roads, while the Road Research Laboratory measurements show that every degree per hundred feet of average curvature on a two-lane rural road reduces average speed by 1.22 m.p.h.

Volume of Traffic

We must now examine how these costs for any given highway are affected by the volume of traffic on it. Clearly, the greater the volume of traffic, the higher will be the average total costs of operation. At low traffic volumes there will be little effect, but as traffic volume rises costs will rise at an increasing rate until conditions of extreme congestion prevail. The problem is how to assess what the levels of costs will be for each volume of traffic.

This problem is conveniently overcome by the A.A.S.H.O.[7] by estimating the volume of traffic that will flow and relating this to the type of highway to give a defined category of congestion for which corrective factors are then applied to vehicle costs. Their method is to divide the thirtieth highest hourly volume by the practical capacity, values up to 0.75 being designated free, 0.75 to 1.25 normal, and 1.25 and over restricted. This suffers from numerous faults. It takes account only of the congestion measured during the thirtieth highest hourly level, levels at other times being disregarded although they may vary widely. Secondly, it relates this to practical capacity, a very nebulous concept defined as "the maximum number of vehicles that can pass a given point on a lane or roadway during one hour under the prevailing roadway and traffic conditions, without unreasonable delay or restriction to the drivers' freedom to maneuver."[8] No attempt is made to define how much delay or restriction is "reasonable," and the "reasonable" amount is not related to the cost of alleviating it. Thirdly, it gives only three categories of congestion, while in practice an unbroken range exists.

A more complete representation of the effect of congestion is achieved by the curve relating vehicle costs to the volume of traffic. The relationship is best assessed by the effect of congestion on speed. The free speed distribution on the road in question can be assessed from measurements on similar roads at times when traffic volumes are low. Vehicle costs related to these speeds will be appropriate for low volumes of traffic. Measurements at different times and on different but similar roads will give data on speed distribution with varying volumes of traffic from very light to very heavy. A comparison of average free speed and average speed with any volume of traffic will give the delay attributable to that volume of traffic, and the time cost of vehicle operation involved. The running costs will be influenced by the spread of the speed distribution as well as its average in two ways. A vehicle travelling at the average speed will not travel at a constant speed, and the more the variation in speed over the trip the higher will be the running costs. The extent of such variations can be assessed by test runs. Vehicles travelling at speeds above the average will have higher running costs, and the difference between the operating costs of fast vehicles and those travelling at average speed will be greater than that between the average and vehicles operating at correspondingly lower speeds. This will be accounted for partly by the extra costs of higher speed, and partly by the greater variations in speed experienced by faster vehicles. The greater the spread around the average speed both in extent and numbers involved, the higher will be average vehicle operating costs. Data on speed distributions can be obtained by traffic surveys, and enough is known about the effects of speed and changes in speed on fuel consumption and so on to assess their effects on vehicle costs.

Users' Personal Costs

This group comprises the costs of highway travel borne by the users personally, as distinct from the costs of vehicle operation. It consists of three types of cost: time, inconvenience, and risk of accidents.

Time

Part of the cost of travel is the time it takes, which time is of value whether it be that of drivers, passengers, pedestrians or goods. Assessing this cost under various conditions calls for two distinct types of calculation; the measurement of the amount of time involved, and the evaluation of it in money terms.

The time cost of goods sent by road is generally of very little importance, but in certain cases it can be a large item. This occurs especially with perishable goods, usually foodstuffs. Here time can be indivisible in the sense that goods which arrive too late for one market by one hour might involve a loss of a further twenty-three hours before the next market. The time of goods in transit is a direct function of distance and average speed and is easy to determine. The value of this time is the interest on the value of the goods plus depreciation. The interest cost will be so small over the differences in time periods with which this analysis is concerned that it can safely be ignored. Where the route is one used by very perishable traffic, however, depreciation may be a significant factor. High-speed motorways in the United States have attracted much of this type of traffic which previously went by rail because of the high time cost of travel on old roads. In this case the time cost of the goods was a more decisive factor than the higher other costs of rail travel.

The time of pedestrians is of some value, but again it will normally be an insignificant amount in the sort of problem with which this analysis deals. Only in exceptional cases will it be worth considering, and its measurement will not then be difficult given the distances involved, average speed of walking, and numbers of pedestrians. Evaluation will be as in the case of vehicle passengers.

Vehicle drivers and passengers are the most important group from a time cost point of view, and the measurement of the amount of time involved for each type of person is not a difficult problem. Traffic surveys and test runs will give the amount of time involved for each type of vehicle at various levels of traffic. For projected roads these figures can be assessed by analogy with comparable existing facilities. A sample survey will give data of vehicle occupancy of different types of vehicles, which will vary with the volume of traffic. At times of traffic congestion buses particularly are likely to be fuller than when traffic is sparse. By such methods it is a straightforward exercise to determine the average numbers of man-hours of each type of person involved in travel over the road in question, for one standard vehicle unit at each level of traffic volume. The much more difficult task now arises of evaluating this in money terms.

It is a generally agreed principle that the value of time of persons travelling in working time is their wage rate. This applies to bus crews, drivers of commercial vehicles, salesmen, and passengers travelling in working time. If we can separate working travellers from non-working travellers the value of time for the former can be determined. This separation is normally achieved by a sample survey. The Social Survey in London in February, 1954, determined details of a week's journeys, subdivided into those in working time and those in leisure time, and incomes for a sample of the population. Correcting week-day figures to allow for week-ends when fewer travellers are in working time it was found that 24 per cent of car occupants, 18 per cent of taxi passengers, and 2.8 per cent of bus passengers were travelling in working time. Combining these figures with the incomes of the groups travelling by each means it was found that the average value of working time per vehicle hour was 49*d.* for private cars, 19.6*d.* for taxis, and 36.6*d.* for buses. These figures apply only to

London and the wage rates are by now out of date. They take no account of the volume of traffic and therefore underestimate average cost since the occupancy of buses for example will be higher the greater the volume of traffic, and their speed will be slower. Thus the working time value per vehicle hour in the case of buses will tend to be higher the slower they travel. A more complete survey along similar lines would give very accurate data of the working time cost of travel. To it must be added drivers' and conductors' wages.

The magnitude of these figures is so great that they can often be the deciding factor in determining the relative efficiences of highway projects. In New York City speeds in peak hours are approximately one half what they are off peak. Half the wage bill of drivers in the peak can therefore be attributed to congestion and the wage bill of drivers on delivery services alone in that city is $700,000,000 per annum. The Road Research Laboratory has calculated[9] that if delays were halved at the ten most congested intersections in central London there would be a saving, at an arbitrary rate of 10/- per vehicle hour, of £500,000 per year. At a rate of interest of 5 per cent this represents a capital value of £10,000,000. It appears from these figures that if those intersections could be improved so as to halve delay, time savings alone would justify £10,000,000 of the cost. The value of time is so important in the economic assessment of highway projects that the wage costs of employed travellers are now included in virtually all highway planning.

The position of non-working time is far less satisfactory. The distinction normally drawn in cost-benefit studies between the values of working time and leisure time and the conclusion that the wage rate can be used for the former but not for the latter rest on false reasoning which deserves close examination in view of their wide acceptance. To arrive at the value of time saved by highway improvements certain assumptions have to be made, and it is in tacitly making inconsistent assumptions that the conventional practice fails. In the simple case of a journey by truck the time of which is reduced from two hours to one hour as a result of highway improvements, one hour of driver's time is saved. At a wage rate of $2.00 per hour the wage cost of the trip is reduced from $4.00 to $2.00. The value of the hour saved, however, depends on what is done with it. If the driver is employed for two hours instead of one and makes two trips, the benefit to the employer is in the form not of one hour's wages saved but of another trip at no extra wage cost. The wage cost of one trip before the improvement was $4.00 and since the employer now gains one trip at no extra cost it could be argued that he is $4.00 better off, not $2.00. If we assume that the number of trips made does not increase there are two possibilities.

If the truck driver works only one hour instead of two, and is paid for only one hour, the employer saves $2.00 but the driver suffers one hour's unemployment. In effect the employer is $2.00 better off, the driver $2.00 worse off, and the driver gains an hours' leisure. Since in money terms the employer and the driver are between them neither better nor worse off, the net result of the improvement is a gain of one hour's leisure time. To value this at the wage rate is to value a gain in leisure time, not a saving in working time. The same argument applies if the driver still received $4.00 for the trip, with no gain to the employer, and gains an hour of leisure time.

Thus to use the wage rate as the value of working time saved, while rejecting it as the value of leisure time, must involve the assumption that the improvement results in more work being done rather than in unemployment. Further, the calculation of the amount of time saved cannot be based on the volume of traffic before the improvement. Nor can it be based on the volume of traffic after the improvement since, although this would give the figure of $4.00 calculated above, it assumes that the extra trip is worth $4.00 to the employer. The trip must be worth less than $4.00 to the employer, as otherwise it would have been made at a wage cost of $4.00 before the improvement. In the absence of precise data on the value of the additional trip to the employer, we could assume that his demand curve for driver's services, showing number of trips made as a function of wage cost per trip, is linear, in which case the calculation of the value of time saved should be on time saved per trip, multiplied by the average of the number of trips before and after the improvement, multiplied by the wage rate.

The assumption that the same number of men are employed for the same amount of time before and after the improvement, that is, that the amount of work done expands in proportion to the reduction in time per trip, is both without justification and contrary to the assumption of a linear demand curve. Without this assumption, however, the use of an average of the number of trips before and after the improvement necessitates an estimate of generated traffic, and also leads to further difficulties.

In a case where the amount of work expands less than in proportion to the reduction in trip time, the use of this technique involves a valuation of leisure time at the wage rate. If, for example, three men were employed to make three trips before the improvement, and two men are employed to make four trips after the improvement, the value of time saved would be one hour per trip times 3.5 (the average of three and four trips) times $2.00, or $7.00. This is composed of $3.00, the value of the extra trip, plus $4.00, the wages of the man no longer employed. But, as has been shown above, this $4.00 is a valuation of the unemployed man's leisure time. To regard it as the value of working time assumes that he can find another job, equally attractive to him, at the same wage rate; it assumes a labour market with vacant jobs waiting to be filled.

The contrary case, when the amount of work expands more than in proportion to the reduction in trip time, would involve the employment of more men, and the use of the existing wage rate would necessitate the assumption that there are men looking for jobs. This, of course, in inconsistent with the above assumption that there are vacant jobs waiting to be filled.

Thus, however we attempt to refine the current practice of valuing working time at the wage rate, we find that it necessitates either the making of unjustified assumptions, or the making of contradictory assumptions, or the valuation of leisure time at the wage rate.

If we assume either that there is no objection to valuing leisure time at the wage rate, or that the labour market is perfect and that the effect on it of the improvement is insignificant (i.e., that any driver unemployed as a result of the improvement can find another job and that any additional drivers required can be found) then the value of working time saved can be conveniently expressed by a formula.

The demand curve for drivers' services can be drawn as in diagram 13 using the amount of work done in one hour before the improvement as the unit. With a constant wage rate per hour, the price of one unit of work varies with time savings; for example, if speeds are doubled, price is halved. The curve is drawn

Diagram 13

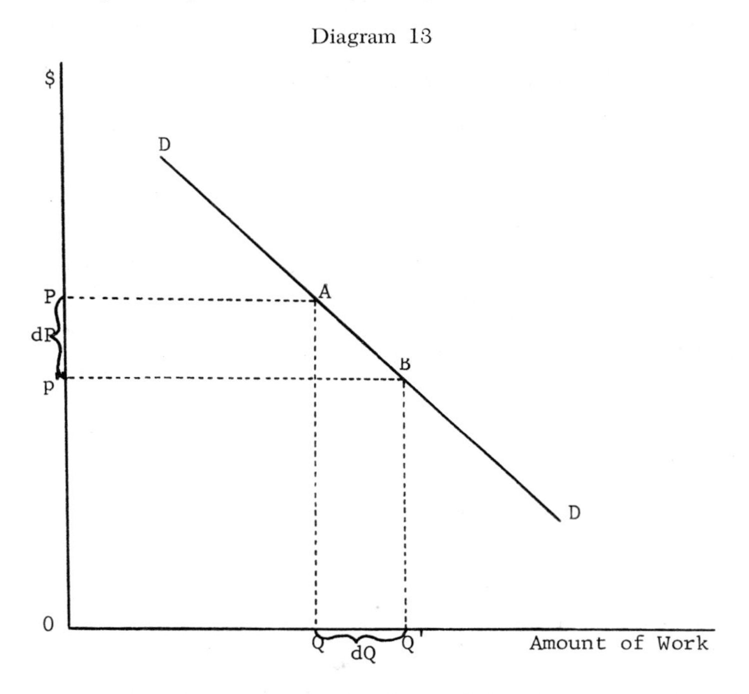

so as to take account of changes in the demand for drivers' services that result from other cost changes, such as vehicle operating costs, which accompany the change in speed.

The position before the improvement in that *OQ* units of service are purchased at a price of *OP*. After the improvement *OQ'* units are purchased at a price of *OP'*. The gain to the employer is the area *PABP'*, being the change in consumer's surplus analogous to our previous concept of net benefit. If we regard the demand curve between *A* and *B* as linear this area is equal to $OQ \times dP + 1/2(dP \times dQ)$ or $dP\ (Q + 1/2\ dQ)$. Elasticity of demand, *E*, is defined as $(P/Q) \cdot (dQ/dP)$. While the elasticity of demand is negative, because *dP* is negative, it is convenient in practice to regard all changes and elasticity as positive. Using the following notation:

T = travelling time in hours per man per day before improvement

M = number of men employed before improvement

W = wage rate per hour

R = proportional reduction in travelling time achieved by improvement

V = value of time saved

Then

$$P = W$$

$$dP = RW$$

$$Q = TM$$

$$dQ = (E.dP.Q)/P, \text{ since } E = (P/Q) . (dQ/dP)$$

$$V = dP\ (Q + 1/2\ dQ) = RW\ (TM + 1/2\ [(E.\ RW.\ TM)/W])$$

Therefore

$$V = RWTM\ (1 + 1/2\ ER)$$

In the case of persons who spend only part of their working time travelling, such as salesmen, a similar technique can be used. Here the demand curve will be for their services performing valuable work, since the time spent travelling between jobs will be of no value to the employer. Using the above notation with the following addition:

S = valuable working time in hours per man per day before improvement

Then

$$P = (W\ [S + T])/S$$

$$dP = (WRT)/S$$

$$Q = SM$$

$$dQ = (E.dP.Q)/P$$

$$V = dP\ (Q + 1/2dQ) = \frac{WRT}{S}\ \left(SM + 1/2 \left\{ \frac{ESM \left(\frac{WRT}{S}\right)}{\frac{W(S + T)}{S}} \right\}\right)$$

Therefore

$$V = RWTM\ (1 + 1/2\ \mathrm{ER}\ [T/(S + T)])$$

Various approaches have been suggested for the evaluation of leisure time spent travelling. The measurement of this relies on the same data as that of working time. Given the total amount of travelling time and the proportion of it during which travellers are employed, the remainder of it is leisure time. The value of leisure time is a subjective matter, closely tied to the comfort of travel, which makes valuation a difficult problem. The peak-time commuter travelling in crowded buses would probably value it very highly if he could have the journey time reduced, the traveller out of peak might have little preference between another ten minutes reading on the bus or another ten minutes reading at home, and some drivers at week-ends might enjoy driving and dislike shorter journey times. Somehow these have to be reconciled. The problem becomes clearer if we separate the time element from the conditions of travel and try first to put a value on time.

One method of attempting this is to collect the traveller's own opinion by a questionnaire. At first sight it appears obvious that the best way to find out how much a person considers his time to be worth is to ask him, but on closer examination many problems arise. On the assumption that a statistically valid sample could be devised, the results of the questionnaire would give four types of answers:

1. Overestimates of time value from those persons who wish to accentuate the case for saving their time. These would probably be few in number.

2. Underestimates from persons who fear that the pricing system on the proposed improvement might be based on benefit. This group would probably be still smaller.

3. Unintentionally inaccurate answers because of failure to think of time as distinct from convenience. If these were combined in the questionnaire the vast majority of people would still find it difficult to think about the marginal money value of their leisure time.

4. Reasonably accurate answers. These would probably be a minority group and could not be identified from the others. Even if they could be identified they would not be reliable as they would come only from persons able and willing to think about the problem analytically. This group would not be a representative sample of all travellers.

For these reasons it appears that a direct attempt to find what value people put on leisure time spent travelling would lead to misleading results.

A more promising approach is the more subtle method of assessing what value people actually put on time in their observed behaviour. Three methods of observation are possible:

1. The route which drivers follow is governed by many factors. If they do not choose the route with the lowest operating costs it must be because either time savings or greater convenience on the route actually used are valued more than the increase in operating costs. An extensive origin, destination, and route study would yield data on this. The extent to which people are prepared to drive further in order to save time or have a more pleasant journey is considerable. A survey on the Shirley Highway, Virginia, showed that 38 per cent of traffic saved distance by use of the freeway but 81 per cent saved time.

The difficulties with using this method are considerable. To collect enough data on which to base any sort of generalization would be an extensive operation. It would give information only about persons able to choose their route, and so would exclude bus passengers. It would confuse time with comfort, as is shown by the fact that many drivers in the United States will drive further, and take more time, to avoid congestion.[10] It assumes that all drivers know the difference in operating cost which they are meeting. Finally, it assumes that data collected in one part of the country will be valid in another.

If we are prepared to accept the imperfections, however, this method has the great advantage of giving a complete schedule. By comparing two routes it can be calculated that a certain amount of time plus convenience costs a certain increment in operating costs. This can be expressed in cents per hour and the survey will show what percentage of travellers consider time worth this price. Other surveys would show what percentage consider time worth a different price. A whole series of such surveys would yield data for a curve showing what proportion of traffic value time at each price Given total anticipated demand these proportions can be converted to numbers of persons to give in effect a demand curve for time savings. Any method which gives this end product is attractive, but the failing must be remembered that by taking account of convenience as well as time it involves comparisons of unlike cases. It cannot be concluded that a person who uses a certain value of extra fuel to save time by going further on a freeway would spend the same amount on tire wear to save the same amount of time over a short cut on a rough track. However, if we limit the survey to cases of extra distances it is reasonable to assume that the only way a longer route can be quicker is by being a "better" road, and the only way to reduce time on the projected improvement is by a "better" road. The data from such a survey would therefore give the travellers' evaluation of "better" roads and include consideration both of time and convenience.

2. A second method is to compare speed with time saved. The operating costs per mile of a vehicle increase with speed, so that if a person drives faster the extra operating cost is the value he places on time saved. Given data on operating costs at all speeds, one can compute the cost of time saved per minute over the trip. This gives in effect a supply curve of time savings, and observed driver behaviour on uncongested roads would give the average demand price, that is, the value the driver puts on time.

This method has many difficulties, both practical and theoretical. It is very rare in many countries that a driver can be observed doing the speed he would like to do if there were no obstacles, and data would therefore be difficult to collect. Even if we could collect data their interpretation in this way rests on three assumptions. Firstly, the driver knows his operating costs at all speeds and is therefore consciously choosing to pay for time. In practice very few people have much idea of the way costs vary with speed. Secondly, it assumes that road conditions are irrelevant. The operating costs for a given speed will vary widely with road and traffic conditions. Thirdly, it assumes that he derives equal pleasure from driving at all speeds. Some people like driving fast and will pay to do so, though the motive is not to save time. Others are more cautious and drive slowly, but would willingly drive faster on safer highways. If we accept these unrealistic assumptions the data are still of limited applicability since their coverage is limited to passenger cars where a person can vary speed, and they give no clue to the value of time where a person cannot drive as fast as he likes, either because of congestion, or because, however valuable his time may be, his vehicle has limitations.

So many factors, therefore, govern the speed at which drivers choose to travel, that to assume their only motive to be willingness to incur increased operating costs to save time, when in practice it is the one they knew least about, is so unrealistic that no reliable conclusions could be drawn from this method.

3. Comparisons with other forms of travel that have different costs and speeds have been suggested as a method of measuring the passengers' willingness to pay to save time. The most common example is the difference between bus and train fares, where the extra train fare is regarded as the value of the time saved. So many other factors such as comfort confuse this method that it is impossible

to isolate time savings. We cannot assume that people are indifferent between rail and road travel at the same speed, still less between road and air travel. Further, the amount of data which could be collected in this way is limited. The most it can tell us is how many people are prepared to pay a certain price to save a certain amount of time; it does not tell us what the value of the time is for any but the marginal rail traveller, nor does it give any clue to the value of time for those travelling by any other means, or where the choice is not available. It is therefore so unreliable and limited a method as to be of no practical value.

One further attempt to assess the value of time by observed behaviour is the use of toll road data. It is found that large numbers of motorists are prepared to pay quite high tolls to travel on first-class highways while lower quality roads are available with no toll. If the toll rate per mile is adjusted for the lower operating costs on motorways, the remainder can be called the price of time saved plus the extra convenience. Again, these two items are difficult to separate and the total value derived from such data would only be usable as the value of time plus the change in conditions from ordinary roads to toll roads. It would have limited applicability to improvements of other types.

Using data on time savings, vehicle cost savings, and toll rates on toll roads in the U.S.A., we find that the price of time plus comfort for passenger cars is approximately 2¢ per minute, or $1.20 per hour. All that we know is that a certain number of persons value time at 2¢ per minute or more. We do not know exactly how much they value it at, nor do we know the value of time to persons not using the toll road, except that it is presumably below 2¢ per minute. Toll road authorities in the United States could give no answers to requests for information about the probable effects of increases or decreases in toll rates on the volume of traffic, and very few changes have in practice been made. It is therefore difficult to assess the demand curve for time except at the price actually charged. In the absence of toll roads in many countries this source of data is not available, and one cannot assume that the value of time is the same in the U.S.A. as elsewhere when converted at current exchange rates.

This comparison of toll and free roads is really a special case of method (1) above, i.e., comparison of different routes where one has higher operating costs but takes less time. As one case in such a survey, toll roads would be useful if available, but many other cases would need to be examined to derive reliable data of general applicability.

In the absence of a generally accepted method of valuing leisure time the normal practice is to assign a purely arbitrary value to it. Highway planners always fear overinvestment in highways rather than underinvestment and therefore tend to err on the side of conservatism in estimating benefits. In the United States this often takes the form of arguing that since one cannot value leisure time precisely it is best ignored. This is really the same as saying that the best estimate available is zero, and when toll roads show large numbers of people prepared to pay considerable amounts to save time, the folly of this convenient way out is manifest. The data are inadequate and unreliable, but since some value has to be put on time in practice, a much closer estimate than zero could be made. The American Association of State Highway Officials adopts a more enlightened practice and recommends, "A value of time for passenger cars of $1.35 per hour, or 2.25 cents per minute . . . as representative of current opinion for a logical and practical value. The typical passenger car has 1.8 persons in it, and a time value of $0.75 per person per hour results in a total of $1.35 per hour."[11] That this arbitrary figure is less than half the wage rate of employed drivers serves to show the tendency to err on the side of conservatism.

One of the simplest ways to value leisure time is to assume that the marginal value of leisure is the wage rate of the person concerned. This has usually been discarded as relying on unjustifiable assumptions, but would be so convenient if valid, that the necessary assumptions deserve further consideration. The basic assumptions are two: that the number of hours spent working is such that the marginal hour's effort is just worth the wage rate, and the next hour has the same marginal value; and that the traveller is indifferent at the margin between work and travel.

Institutional obstacles are normally held to invalidate the assumption that a person works the number of hours he chooses at the standard wage rate. If a forty-hour week is standard in the firm the individual cannot normally work thirty-nine hours or forty-one hours. Some persons, mainly professional and self-employed, are free from these restrictions, but for the majority they are very real. The question then arises why the firm works a forty-hour week. With the current state of organization in the labour market it is surely not unreasonable to assume that if all employees in a firm would prefer a longer or shorter working week this would be achieved, with a compensating decrease or increase in the numbers employed. From this it follows that the number of hours actually worked in some way reflects the average desires of the employees. Some will value the last hour's time above the wage rate and would willingly sacrifice an hour's wages for an extra hour's leisure; others would willingly work another hour for another hour's pay. But if the number of hours worked is an average then on balance these will cancel out. We can therefore conclude that the average person values his last hour's work at the wage rate.

The marginal value of time to the individual will be a continuous curve, and the values of two adjacent units will not differ widely. If they differ at all the next hour, that is, the marginal leisure hour, will be worth more, and this is reflected in the fact that overtime rates of pay are above the standard rates. The difference between standard and overtime rates will be greater than that between the value of adjacent hours, since the overtime rate will be the marginal value of overtime and will normally allow for more than an hour of overtime per week. Where a person normally works overtime the overtime rate will be the marginal value of leisure. Our first assumption is therefore quite a realistic one when used as an average to apply to all employed persons.

The second assumption, that the traveller is indifferent at the margin between work and travel, causes more difficulty. Certainly some people are less comfortable and less happy in crowded peak-hour public transport than in their offices, but the same does not apply to pleasure travel at week-ends or on holiday. Indeed if a person were indifferent between a pleasure drive on a Sunday afternoon and the same time spent at work, he would presumably be at work. We must therefore distinguish between travel which is an undesirable necessity to transport oneself from one point to another, and travel which is a pleasant relaxation. The majority of passenger hours are of the former type, however, and these occur predominantly at peak travelling times. The greatest savings in time, both in amount and numbers, achieved by a highway improvement will be in the most congested periods, and to value all time at the rate applicable to unpleasant travel would be a closer approximation than to adopt the opposite standard. Indeed, even pleasure travel at week-ends is predominantly undertaken to get somewhere, rather than merely to travel, and here again time will have a positive value. Some people may well value a reduction of twenty-five minutes each way in their weekly trip to the cottage at more than a five-minute reduction each way, five times a week to and from work. This is reflected partly in the growing tendency to live further from the centres of large cities. It therefore seems reasonable to assume that on the average the value of travel time will not vary greatly over the vast bulk of trips with the purpose of the trip, the pure joy-ride and the absolute emergency being relatively unimportant and then counterbalancing. The value of time saved will mainly be a function of the comfort or otherwise of travel. The really controversial assumption therefore is that on the average travel is as unpleasant as work.

This obstacle can be overcome by using a correction factor. If, on the average, travelling is half as unpleasant as work, i.e., a person is indifferent between one hour's work and two hours' travel, the correction factor is 0.5 and the value of travel time is 0.5 times the wage rate. If on the average they are equally unpleasant, the correction factor is 1. This method of valuing leisure time is therefore simple, practical, and reasonably accurate if we can determine the value of the correction factor. This might well be done by an extensive survey on the lines outlined above to determine the expense people will voluntarily incur to save travelling time, and relating this to the average wage rate to derive the value of the correction factor. It seems probable that with diverse traffic the value of the correction factor will not vary greatly between proposed improvements, and a figure based on averages would therefore be satisfactory.

The problem of the relative inconvenience of work and travel also applies to the valuation of working time. Truck drivers will not be indifferent to highway conditions, and the attractiveness of their work is therefore related to road and traffic conditions. A driver might well prefer to cover a greater mileage on a modern highway than a lesser mileage in congested conditions in the same time. The attractiveness of the job operates through the labour market to affect wage rates, and these are often different between urban and rural areas. If an improvement affects the attractiveness of a driver's job sufficiently to have an effect on his wage rate, this will be reflected by valuing time on each road at the appropriate wage rate. The same considerations apply to those such as salesmen who spend only part of their working time travelling. Some might prefer a situation where a larger part of their working day is spent travelling rather than working; others might prefer the converse, especially if they work on a commission basis. Where the average is sufficiently strong in one direction to have an effect on the wage rate, it will be allowed for in our analysis by using an appropriate wage rate on each road under analysis.

Thus, by using appropriate wage rates in each case for working time, we can take account both of time and inconvenience. Similarly by valuing non-working time at the wage rate, modified by a correction factor to take account of the relative inconvenience of work and travel, the analysis takes account of both the time and inconvenience of travel. No separate allowance for the cost of inconvenience is needed. A person's normal work is taken as the standard level of inconvenience, and at this level the value of time is the wage rate. The intensity of the inconvenience of travel determines the correction factor while the quantity of inconvenience is proportional to the time it lasts.

While we have examined above the techniques of valuing time saved by an improvement, because this is the approach used in conventional cost-benefit analysis, it must be emphasized that the analysis outlined in previous chapters does not attempt to measure this directly. We are concerned not with the value of time saved but with the cost of time used. In drawing the user's personal cost curve the amount of time for each person at any volume of traffic is determined by the length of the trip and average speed. The value of time is the wage rate, corrected for non-working time by a factor which takes account of inconvenience. Where data are not available to assess the size of the correction factor it can be ignored. Problems of generated traffic, which as we saw above cause considerable difficulty when trying to assess the value of time saved by cost-benefit analysis, do not enter into the cost of time used in our analysis. Both the amount of generated traffic and its value are represented by the demand curve.

Accidents

A reduction in accident rates resulting from highway improvements can bring about savings in two ways, both of which make this an important consideration in deciding whether a proposed improvement is worth while. The direct costs of death, suffering, and lost output associated with accidents will be reduced, and if the over-all improvement in national figures is sufficiently great this will lead after a time to a reduction in insurance rates.

As in the case of time costs the assessment of accurate data of accident costs is a difficult problem; and again falls into two parts which can be examined in turn. These are the numbers of accidents of different types and the evaluation of them in money terms.

Anticipation of the number of accidents that will occur on a projected highway involves two major difficulties. Severity of accidents is difficult to classify because they vary over a continuous range, and the numbers involved on a single highway are so small that probability becomes important. The only distinction usually drawn in accident rate statistics is between those which involve injuries or fatalities and those which do not. This distinction is useful for evaluation purposes but does not demark distinct degrees of severity. The probability element can be overcome only by using average data and assuming that the projected highway will have average accident rates for highways of its type. This might mean in a particular case one fatality every two years, while in practice there might be two in a week or none for five years. This does not invalidate the use of averages, however, as they remain the most reliable estimate that can be given, and errors will necessarily counterbalance. The principle of all insurance is that similar cases with the same risks turn out to differ widely in practice, and the use of the same accident cost curve for similar highways, even though they may turn out in practice to have widely different accident records over short periods of time, can be regarded as an appropriate use of the principle of insurance. An improvement which lessens the risk of accidents is not neces-

TABLE IV

ACCIDENT RATES ON DIFFERENT TYPES OF ROADS

Type of highway	Traffic volume: vehicles per day	Accidents per 100 million vehicle miles: Fatal	Total
Two-lane highways			
U.S. average	Under 5,000		350
	5,000-9,000		450
	9,000 and over		260
	All volume average		370
Three-lane highways			
U.S. average	Under 5,000		380
	5,000-9,000		410
	9,000 and over		1110
	All volume average		500
Four-lane undivided highways			
U.S. average	Under 5,000		230
	5,000-9,000		400
	9,000 and over		380
	All volume average		370
Four-lane divided highways			
U.S. average	5,000-9,000		320
	9,000 and over		410
	All volume average		380
Specific surface streets			
Los Angeles:			
Figueroa St. & Wilshire Blvd.		4.2	260
London:			
Main streets in Central London		8.0	600
Great West Road (A.4)		38.0	690
Uxbridge Road (A.4020)		43.0	710
Limited access freeways:			
San Francisco Bayshore Freeway	83,000	2.2	
Hollywood Freeway	79,000	1.8	
N.Y. Metropolitan systems	59,000	1.7	
Shirley Hwy., Arlington Co., Va.	32,000	1.7	
Wilbur Cross Hwy., Conn.	15,000	1.3	
Dallas Central Expressway	36,000	0.0	
Kansas City, S.W. Trafficway	30,000	0.0	
Washington, D.C., Whitehurst Freeway	28,000	0.0	
Comparative U.S. national averages			
Expressways		2.8	179
Other roads with similar traffic		10.1	425

SOURCES: David M. Baldwin, "The relation of highway design to traffic accident experience," Convention Group Meetings, A.A.S.H.O., 1946; Lloyd Aldrich, "Comparison of parkways with surface streets in capacity, time savings and safety" (Los Angeles, 1952); W. H. Glanville and J. F. A. Baker, "Urban Motorways in Great Britain," B.R.F., Urban Motorways Conference, 1956; J. Barnett, "Urban Motorway Development in the United States," B.R.F., Urban Motorways Conference, 1956; A.S. Hadgliss, "Economic Justification for Urban Motorways," B.R.F., Urban Motorways Conference, 1956.

NOTE: The degree of access control on limited access expressways varies considerably. These examples have full control of access. Those with lower standards might have higher accident rates.

sarily a failure if the accident record does not improve immediately, any more than a fire escape is an extravagance if in practice there is never a fire.

Some data are available of accident rates on different types of roads under different conditions, but they are generally inadequate and also unreliable in that comparisons are often made between unlike cases. In the data above, for example, no account is taken of any other factor than the number of lanes, though many other things are important and may vary. Over large samples these other factors tend to cancel out, but care must be taken in drawing unreliable conclusions from small samples. A more complete recording and classification of accidents is needed to overcome this. Some general conclusions can be drawn from available data, however, of the relationship of highway type to accident rates. The data in table IV include examples of various types of highway with different traffic volumes and accident rates.

From these data it appears that two-lane and four-lane undivided roads have a peak in the accident rate at traffic volumes between 5.000 and 9,000 vehicles per day, with a higher volume for four-lane than for two-lane. Four-lane divided highways also have a peak but it is at over 9,000 vehicles per day. The probable reason for this peak is that the risk of accidents increases with the volume of traffic but falls with reduced speed. At volumes above the peak accident rate, the reduction in speed caused by congestion more than outweighs the extra traffic. Congestion and speed also have an effect on the severity of accidents. A sample of police data for 1935–38 shows that 24 per cent of injury accidents in light traffic, but only 19 per cent in dense traffic, involved serious injury. Twenty per cent of injury accidents occuring at a combined speed below 20 m.p.h. involved serious injury, compared with 30 per cent at combined speeds above 20 m.p.h.[12]

Three-lane highways show a marked increase in the accident rate at high traffic volumes, due to opposing streams overtaking in the same lane. The effect of the mode of contact on the severity of accidents is shown in a sample of police data for 1935-38. Twenty-six percent of head-on injury accidents were serious, but only 15 per cent of side-to-side, and 11 per cent of head to tail. The side-to-side figure includes vehicles travelling in both the same and opposite directions. Avoidance of head-on and side-to-side accidents between vehicles travelling in opposite directions accounts for the lower accident rate on dual carriageways than on three-lane roads at high traffic volumes.[13]

Surface streets generally have much higher accident rates than limited-access freeways, but after a point congestion causes a reduction in accidents and a marked reduction in fatalities probably due to reduced speed. Freeways show a tendency to increased fatality rates with increased volumes of traffic, but individual roads differ so much in standards of safety that it is dangerous to draw conclusions from limited data. The effect of differing degrees of access control and of different numbers of roadside features are shown in tables V and VI. Although the accident rate appears higher the more built up the area, and higher in urban than rural areas, the fatality rate is higher in rural areas, due to the greater severity of accidents at higher speed. In 1953 the U.S. average fatality rate per 100 m.v.m. was 9.3 in rural areas and 4.5 in urban areas, the over-all figure being 7.1. The greater severity of accidents in non-built-up areas is shown in English police data for 1935-38. Twenty-two per cent of injury ac-

TABLE V

SAMPLE SURVEY OF U.S. CONTROLLED-ACCESS HIGHWAYS

Degree of access control	*Miles in sample*	*Accidents per 100 m.v.m* Fatal	Total
None	899	8.0	408
Partial	849	9.6	240
Full	434	2.8	171

SOURCE: J. Barnett, "Urban Motorway Development in the United States," B.R.F., Urban Motorways Conference, 1956.

TABLE VI

SURVEY OF EFFECT OF ROADSIDE FEATURES IN MICHIGAN IN 1952

Percentage of Area Studied	Number of roadside features per 1000 ft.	Accidents per 100 m.v.m.
28.0	None	374
49.8	1 - 3.99	906
22.2	4 & over	1348

SOURCE: *Controlled Access Highways,* Kansas Legislative Council Research Department, Publication no. 182, 1953.

cidents in built-up areas were serious compared with 30 per cent in non-built-up areas. A Ministry of Transport census for 1936-37 showed 25 per cent and 40 per cent respectively.[14]

Many other features of road design affect accidents rates, one of the most important of which is alignment. This is clearly shown by the results in table VII of a survey of accident rates on curves on two-lane rural roads in Buckinghamshire, England, made by the Road Research Laboratory in 1952.

TABLE VII

EFFECT OF CURVATURE ON ACCIDENT RATES

Curvature in degrees per 100 ft.	Personal injury accidents per 100 m.v.m.
0.0 - 1.9	250
2.0 - 3.9	300
4.0 - 5.9	350
6.0 - 9.9	380
10.0 - 14.9	1,360
15.0 & over	1,490

SOURCE: T. M. Coburn, "Accident, speed and layout data on rural roads in Buckinghamshire," Department of Scientific and Industrial Research, Road Research Laboratory, 1952.

The greater severity of accidents on bends is shown also by police data for 1935-38. Thirty-two per cent of injury accidents on bends were serious compared with 23 per cent on straight roads. A Ministry of Transport census for 1936-37 showed 38 per cent and 30 per cent respectively.[15]

Data are not available of the precise effects of other factors on accident rates, and further research is needed to collect them. Many features will be important, such as gradient, sight distance, surface, speed; and the ideal would be to have sufficient data to classify a projected highway according to its design, and predict its accident rate as the average for its class. Though much work remains before this can be done, it is by no means an impossible task and could be achieved in stages of increasing accuracy. The estimated accident rates on the projected highway at various traffic volumes, must then be converted to money terms.

The costs of accidents can be divided into: property damage, damage to vehicles, medical expenses, temporary and permanent incapacity, administrative and legal costs, personal cost of injury, and death. Property and vehicle damage, including the losses during periods of repair, can be determined from repair and replacement costs. Payments by insurance companies are a possible source of data. Medical expenses resulting from injury can likewise be valued. Temporary and permanent incapacity are meant to include only loss of earning power; the personal inconvenience is a personal cost of injury. Again the payments for compensation by insurance companies reflect the cost of such incapacity. Compensation often combines loss of earning power with personal cost but these can usually be separated.

Working on these lines Professor Jones estimated the total cost to Britain of road accidents for the period 1935-38 at approximately £60 million per annun or 1.33 per cent of the national income.[16] Reynolds estimated the cost for 1952 at £72 million.[17] Adjusting Jones's estimate for changes in the value of money would give £110 million for 1952, which is considerably higher than Reynolds' estimate. Seven million pounds of this is accounted for by a fall in numbers of casualties and fatalities between 1938 and 1952, and the remainder results largely from a different basis of calculation. By taking compensation paid by the courts as a basis, Jones's estimates include a measure of compensation for personal suffering, which is specifically excluded by Reynolds by estimating the costs of injury and death on the basis of average output and consumption figures for the age groups concerned.

Taking into account only the real cost of accidents to the community in terms of monetary expenses and lost output, Reynolds estimated the average cost of accidents in 1952 as shown in table VIII.

The great limitation of both Jones's and Reynolds' estimates is that they take no account of the personal costs of pain and suffering associated with injury and death, and the perpetual fear of accident of all who use the roads. Not only are these figures therefore incomplete, but taken by themselves they are positively misleading. This is particularly true of the cost of death, which Reynolds calculates as the present value of the sacrificed expected output over the life expectancy of the person killed, less his expected consumption of goods and services. Retired people have no expected output but some consumption, while the useful work of unemployed women does not feature in the statistics. As a result the cost of death of men over sixty and women over thirty is negative. We could therefore reduce the total cost of accidents on Reynolds' reckoning simply by killing off more women and old men. It is true that the community would be better off financially if old people were killed, but the community still regards

TABLE VIII

COST OF ROAD ACCIDENTS

	Average cost £
Death:	
All persons	2,000
Children (under 15)	3,000
Adults	1,800
Injury:	
Serious	520
Slight	40
Cost of damage to property per accident (injury and non-injury)	32
Administrative Costs:	
Per injury accident	56
Per non-injury accident	6
Total cost per injury accident	332
Total cost per non-injury accident	38

SOURCE: D. J. Reynolds, "The Cost of Road Accidents," *Journal of the Royal Statistical Society*, vol. 119, part IV, 1956, p. 403.

such deaths in road accidents as a loss. An economic assessment of the full costs of road accidents in terms of what it would be worth to avoid them must therefore include a valuation of human life quite apart from what that life could produce in material terms. Some allowance must also be made for pain and suffering which do not involve financial cost.

These items of the personal cost of injury and death are difficult to value. Although very much a subjective matter, they are impossible to value subjectively. The task of valuing the personal cost of injury is shouldered by the courts in assessing compensation, and in the absence of any better source these data can be used as the value society puts on injury. The value of life itself, however, is not so easy. A person injured in a road accident claims both for loss of earnings and for personal hardship. If he is killed, the next of kin claim for their financial interest in his potential earnings, but no one claims for the lost life. If we use compensation figures only therefore, injury might cost more than death.

The actual value which society puts on human life varies widely and depends mainly on the amount of sentiment aroused by the way it is lost. If a child is missing no expense is spared in the effort to find him and save his life, but if the same amount of money spent on road improvements would improve an accident black spot so as to save two unknown childrens' lives every year it is begrudged. If one person is killed in an air crash it is the object of a full inquiry; if a thousand are killed on the roads it is a matter of course. The amount of money spent on railways to increase safety is greatly different per expected life saved from that spent on roads. It is perhaps too much to expect society to be sufficiently consistent to apply a common scale of value to life on all occasions, and there is justification for some differences. Many persons killed on the roads are partly to blame for their death, while very few killed on railways are responsible. The latter might be regarded as the greater tragedy, and worth a greater amount of money to avoid. But there is no reason why a common scale of value should not be applied to similar cases. At present the decision whether a certain highway improvement, expected to bring about a certain reduction in accidents, is worth the cost is a political one. The person responsible has to put a value on human life. The inconsistency arises where this value differs between cases. Available funds could be used more efficiently if a scale of value of life saved were laid down, and the highway planner used this as one basis of economic assessments. The National Safety Council in the U.S.A. lays down the following schedule, shown in table IX, use of which is recommended by the A.A.S.H.O.

TABLE IX

COST OF DEATH IN ROAD ACCIDENTS

Age	Cost of death per person Male	Female
0 - 14 years	$17,000	$ 8,000
15 - 55 years	29,000	17,000
56 years and older	5,000	3,500

Personal injuries, average cost $660 each.
Property damage accidents over $25 average $160 each.

A similar schedule for the cost of death could be laid down for other countries by the ministers politically responsible for expressing the opinions of society in this field, and the average cost of injury and property damage can be assessed from data of compensation paid by insurance companies, as has been done by Jones and Reynolds. Combining this schedule with the expected accident rates would give the accident cost curve which is one of the components of the total users' personal cost curve.

COMMUNITY COSTS

These are the items of cost and benefit which accrue as a result of highway improvements to persons not using the highway, or to the community as a whole. They are mainly indirectly related to the primary function of the highway and no account of them would normally be taken by private enterprise. Railway companies, for example, did not count smoke and noise nuisance to neighbouring properties as a cost. The current trend of opinion is to pay more attention to such matters, however, by means of development controls, town and country planning, smokeless zones, and so on, and a complete analysis which claims to determine the optimum plan must take account of the advantages and disadvantages of each, however indirect, to whatever extent this is practically possible. No account must be taken, however, of some indirect results which are not benefits or costs of the highway, but which often confuse thinking on this subject. Thus, if highway development results in railways making losses, these losses are in no sense costs of the highway. They may result from railways pricing themselves out of the market by trying to recoup historic costs, in which case they are the result of misguided policy by the railways. If highway improvements force reductions in railway rates but the railways keep the traffic, there is a shift in the incidence of benefit from the railways to the users, but none of the benefit is lost in the shift. If as a result railways cannot even cover prime costs then highways are fulfilling a service for less than it would otherwise cost, and the net effect is a saving, not a cost. There is similarly no foundation for the argument that by attributing the consumers' surplus on diverted traffic to the highway we are double-counting the benefit otherwise obtainable on the railway. The demand curve for highway use will have a ceiling at the rate at which traffic would change to rail. Net benefit derived from highways is the area above average total cost and under the demand curve for highway travel, not that for travel by any means, and therefore includes only the additional benefit over what would be derived from the most competitive alternative form of transport.

Certain new locations for highways may be of great value in necessitating the clearing of slum areas; others may spoil the natural beauty of the countryside. Allowance for these aspects is partly included in the cost of right of way. The cost of a route which involves destruction of buildings already due for demolition will be lower than one through better property. Land owned by the nation as a beauty spot may well be of higher value than other undeveloped land. Where highway location coincides with a proposed subsidized slum clearance scheme the amount by which the highway construction reduces the cost of the scheme is an additional negative community cost. Where a highway spoils the beauty of a region for the community, even though the land was privately owned and of little economic value, there is an additional cost to the community. This could be taken into account by a scheme of development charges on highways which would put a higher value on the use of land the greater the inconvenience to the community involved. The great advantage of taking such matters into account in the planning analysis, even if this is by the arbitrary opinion of the planning authority with no financial compensation, is that when all other costs of two alternative locations are equal but one is the greater eyesore, the other will be chosen. The amount allowed for such items will be important where other costs are slightly in favour of the greater eyesore, as they will then be the deciding factor. No guidance can be given as to the method of evaluation, as it is purely a matter of opinion, and in the last resort must be left to the decision of the appropriate minister acting as the representative of the people. The aim should be to determine how much it is worth to the community to avoid having a highway on the site in question, and this can be regarded as the community cost of having it.

The existence of a new or improved highway will have an effect on some traffic which does not use it. This will mainly be beneficial by reducing congestion on roads from which some traffic will be diverted. The extent of this will not be difficult to value, given the cost curves for the relieved roads, excluding irrelevant historic costs, and the expected change in traffic volumes on them. Where a limited access highway forms a barrier, however, some traffic crossing it may be put to extra cost by a diversion. The cost of this will depend on the amount of such traffic and the extra distance and time involved.

The value of improved highways in facilitating educational and recreational travel is often stressed as a community benefit. It applies only to those who travel, however, and is therefore included in the demand pattern.

When highway work is undertaken in times and places of unemployment there is a definite community benefit. This could be allowed for by excluding the cost of labour otherwise idle from the construction costs, or costing it only at the difference in costs of idle and occupied labour. In practice it will be more convenient, however, to include the full labour costs as a cost of construction, and the amount saved by reduced unemployment relief as a negative community cost. It is not practical to allow for the reduced psychological costs of idleness as distinct from the financial relief.

It is often argued as a point in favour of good highways that they are of vital military importance in wartime. In practice this is a two-sided advantage, as was learned by German experience, the autobahns being as useful to the allies in invading Germany is they were to the Germans in invading the rest of Europe. Nevertheless, this is an important item of community benefit, or negative community cost, which should be included. It is best valued by the defence departments by their normal methods. The relative value of good roads and modern fighters is a similar problem to the relative value of modern fighters and battleships. The defence departments should therefore be asked to estimate the value of proposed major highways, knowing that their estimate will play its part in the final decision.

Perhaps the most apparent item of community cost, and that often considered the most important, is the effect of highway developments on land values in the area. Changes in the access to property and in the volume and pattern of traffic can have a considerable effect on the value of a site. This, however, can only be due to changes in traffic on the highway, or anticipations of it. The traffic on a highway after an improvement will consist of four groups; previously existing traffic, diverted traffic, generated traffic, and secular growth in traffic. What effects will each of these have on land values?

The traffic which continues to flow after the improvement just as before will bring about no changes. Diverted traffic will carry all its attributes to its new location. If traffic which used to bring trade to businesses along its route moves to the new route, it will take its trade with it. If the new road is of limited access design there will be little if any roadside business, and traffic for which business was an important part of the trip will not move. This will not apply where new business properties develop along the new route. In any case it can be assumed that the aggregate volume of trade will not be affected by the diversion of traffic. Where trade is diverted there will be a tendency for property values in the new location to rise and those in the old location to fall. Since the total volume of trade is unaffected these will tend to be equal. The net loss resulting from the project is therefore the cost of moving business premises to the new location.

Traffic may be generated on both the new and old routes. Generated traffic on the new route will tend to increase trade just as did diverted traffic. The effect of traffic diverted from the old road will be to reduce congestion on it, which may well attract new traffic. The traffic diverted will tend to be that for which trade was not an important part of the trip, while the traffic generated on the old road will tend to be that for which it is important; otherwise it would have been generated on the new road. The net effect on the old road may well be that while the volume of traffic is smaller a larger proportion of it brings trade, which is of greater total value than before. On the new road, businesses patronized by all traffic, such as service stations and cafés, will increase most, especially if the new road has access control.

Thus, while property values along the new location will increase, especially if access is not controlled, this will not be at the expense of the old location where trade from new traffic might well outweight that lost by diversion. The generated traffic on the old road will bring diverted trade, however, and other business centres further from the centre of town may therefore lose trade, and property values fall. These results are borne out by American experience when bypasses or throughways divert traffic from the business centres of towns. The through traffic which has been diverted brought very little trade, but its removal eases congestion and traffic flows in from the suburbs for business. The net effect is that property values rise most in the downtown centre which has been bypassed and fall in suburban business districts. When the new road has no access control, property values rise more along it and less in the business centre. If the total volume of trade is not affected we can expect the increases and decreases in property values to balance out, leaving a net community cost of relocating businesses whose trade has fallen as the result of the improvement. These costs may well be partly counterbalanced by the effect of increased trade brought by generated traffic and secular growth. The gains from reduced transport costs will be spent by the community partly on more transport and partly on other goods. The latter will bring increased total trade, but the extent of this as the result of a single improvement will be very small. In periods of growth in economic activity business will be increasing. The relocation brought about by the highway improvement will then cause some areas to expand faster than others, but owing to the overall rise it is possible that none will decline. Disturbance costs will then be avoided.

The conclusion as far as business premises are concerned is that highway improvements bring little if any increase in total trade, though nation-wide improvements bring increased over-all prosperity in the long run; and since the value of business premises depends largely on the volume of trade there will be little net change. Property values will be affected favourably in some areas and unfavourably in others, and the extent of these differences will be less if the new facility has limited access. Some landlords will be better off and some worse off, but the net community cost will be only the disturbance costs of the businesses whose trade falls as a result of the improvement to such an extent that they move to a different location. This will be less in a period of growing trade.

Reduced transport costs and improved access to agricultural and industrial sites will tend to increase their values. This is a shift in the incidence of benefits to traffic, however, and not a separate item of benefit. Traffic that flowed at the previous higher costs will now move at the lower cost. This gain is included in the benefits to traffic, and where it is not completely removed by increased taxation some of the remaining benefit will be capitalized in site values. Where new traffic is generated, say where a new road gives access to a site hitherto unsuitable for development, the gain will be the difference between the demand price for such traffic and its cost. Again this is included as a benefit to traffic. That no separate benefit accrues to the site can be seen from the fact that if the surplus were entirely removed by increased taxation the traffic would not flow and the site would not be developed. Any increase in the value of the site can therefore come only from a shifting of the reduction in transport costs, and must be included in the cost and demand curves. The same arguments apply where improved roads bring residential development further from city centres.

The impact of road developments on the pattern of a community might be considerable. The cost of travel to work is reduced in terms of both money and time, and people can therefore live further from their place of work. This gives people living in a given place a wider possible area in which to seek employment, and a given firm a wider area from which to draw workers. As a result people live in more desirable locations and find more suitable jobs. The benefits derived by the community from such developments are composed entirely of the benefits derived from each individual trip. If more desirable employment is available in a more distant town a person might consider the journey just worth a certain amount. If the cost is brought below this the net benefit is the difference between this demand price and the actual cost. This might be divided between employer and employee by the payment of wages less than the employer would have been prepared to pay to secure this person, the lower rate being brought about by other suitable candidates also coming within range; but, however divided, the total net benefit is the difference between what the trip is worth and what it costs, no more and no less.

Land values arise only from scarcity of certain types, and location is very often the important scarce factor. Improved transport facilities will reduce the relative location advantage of already developed sites by competition from those acquiring improved access. The more good roads that are built the less will be the differences in locational advantage. In the long run, therefore, the effect of good roads might possibly be to reduce aggregate land values. If changes in land values are included as costs and benefits this effect will be reflected as a community cost, while in fact it is a community gain. This can be illustrated by an analogy. If climatic changes suddenly converted deserts to fertile agricultural land to such an extent that there was more fertile land than anyone wanted, and there were no transport costs, agricultural rents would fall to zero and total land values fall. Such a change would be beneficial, not harmful, to humanity. Similarly, if reductions in transport costs reduce the scarcity of land with good location, they are a community benefit. The extent of this benefit will always be reflected by measuring the benefits to traffic, but not by measuring the shifted benefits to land values.

We therefore conclude that, though changes in land values may be a useful point at which to levy highway taxes, they can never be a separate item of cost or benefit from the changes in transport costs which brought them about, and can grossly misrepresent the extent of those changes. The only community cost remaining as a result of relocation brought about by highway developments is the actual disturbance cost of moving businesses from which trade has been diverted.

DEMAND

We have examined in this chapter the components, of, and methods of measuring and evaluating, all the costs outlined in chapter II. It remains to examine the method by which the demand curves used in chapter III can be estimated. This is done partly by measurement of existing conditions, and partly by applying the lessons of experience on other projects. The four sources of traffic pose different problems.

Existing Traffic

Where an existing road is being improved as distinct from a completely new road built, the existing traffic will continue to flow at reduced costs. The amount is easily measured by a traffic survey. The effective demand curve for the new road will be a horizontal straight line for this volume at a level of cost equal to that existing on the road before the improvement. Demand price might be higher at lower volumes, but since this part of the curve will be used in measuring only the net benefit of the improvement, we must exclude the benefit which already existed. Thus if the demand curve for travel over the route is DD^1 in diagram 14 and the operating cost level before improvement *OA*, existing traffic volume will be *OC*. As a result of the improvement the cost level is reduced from *OA* to *OB*, and *CE* extra traffic flows. The net benefit derived by *OC* existing traffic from the improvement is the reduction in cost per unit, *AB*, multiplied by the number of vehicle units *OC*, i.e., *BGFA*. The area *FGH* will be net

Diagram 14

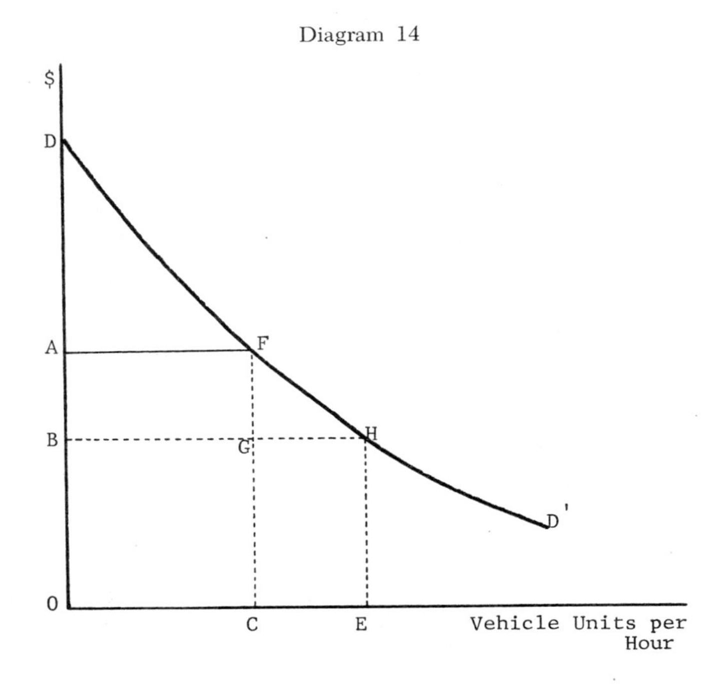

benefit to generated traffic. Thus net benefit is measured by the area under the demand curve, less costs, if we regard the demand curve as $AFHD^1$. This flat top to the demand curve will not apply in the case of a new road on a new location where there is no existing traffic.

Diverted Traffic

The amount of traffic which will be diverted to the new road is a function of two things; the suitability of the new route to the origins and destinations of traffic, and the relative cost levels on the new road and other roads. The first requirement, therefore, is an origin and destination study of traffic in the area which might be diverted to the new road. There are various methods of conducting origin and destination surveys.[18] The direct interview method determines origins, destinations, and stops of a sample of vehicles passing a given point. It is most suitable for light traffic volumes on main roads in rural areas. Where volumes are heavy or the road layout complex and we therefore require further information, the postcard method can be used. A questionnaire on a postcard is handed to each driver passing the check point with a request to fill it in and mail it. Where a complicated network of roads is involved, a ring of check points can be put round it and vehicle entries and exits recorded with times, either by tagging vehicles or recording registration numbers. A full survey of movements in a densely populated area can be obtained by interviewing a sample of persons at home to determine the trips undertaken by members of the family in a given period, usually a week.

Given the data on traffic movements on the old roads, and the costs on them as determined by test runs, we can estimate the demand on the new road from diverted traffic. It is assumed that traffic will be indifferent between two routes with the same total operating costs, including time, convenience, and so on. Some traffic will be able to use the new road by a direct link with its existing route, some will be able to use it with a detour, and some will not use it at all because it does not go towards their destination. The demand price of traffic which can use it without a detour is the prevailing cost on the road at present used. Where a detour is involved the demand price is the existing cost less the operating costs on the detour links. With the data of the routes and costs of all traffic at present, a curve can be constructed to show what volumes will divert to the new road at each cost level.

Where diversion from an existing road is so great as to reduce cost levels on it, this must be allowed for in the demand curve for the new road. The first vehicle will move at any cost up to the level prevailing on the old road. As more move, the cost on the old road will fall and the demand curve for the new road will likewise slope down to the right. The case where this is most likely to happen is that of a bypass or throughway diverting traffic from a congested area. Here diversion may be entirely from the old through road to the new bypass. The maximum traffic which might use the bypass is the volume of through traffic. Analysis of data collected in the United States and Germany shows that there is an inverse relationship between population and the proportion of through traffic, and this is surprisingly consistent in different cases.[19] The through traffic is the maximum which could be diverted, and this will move if the cost level on the new road is below the level on the old road when relieved of this traffic. Lesser volumes will divert at higher cost levels, the total operating costs on the two roads remaining equal, until the level of cost on the new road is as high as that on the old road before any diversion, when none will move. The cost levels on the two roads will be affected by generated traffic. Thus, if the old road attracts new traffic as a result of relieved congestion, operating costs on it will be higher than otherwise, and more traffic will divert to the new road.

Generated Traffic

The extent to which reduced cost levels will generate new traffic is difficult to estimate. It depends on the elasticity of demand for movement on the existing road and roads from which traffic might be diverted. This in turn depends on population, incomes, business and social relationships, and travel habits in the area. Some estimation of generated traffic must be made, however, and this is best done in the light of experience with other comparable facilities. Extensive empirical studies have attempted to relate the amount of travel between two points to population and distance.[20] This relationship can be expressed by the formula, $T = (K.P1.P2.)/D^n$, where:

T = Traffic flow per unit of time between two areas.

$P1, P2$ = Populations of the two areas.

D = Distance between the two areas. This can be measured in terms of travel cost, since a time reduction is as important as a reduction in mileage.

K = A constant, dependent on incomes, etc.

n = A positive exponent found empirically to vary between 0.6 and 3.36.

Further research is needed to determine the exact components of K and n, and their relationship, and until this is finally settled the use of figures found appropriate in similar cases will give the most accurate estimate possible. Where effective distance is reduced by dD the amount of the increase in total traffic will be $T([D/(D-dD)]^n-1)$. Given the volume of traffic and level of cost before the improvement, and the value of n, this formula can be expressed as a demand curve for generated traffic on the new road at cost levels below that existing before the improvement, the volume of generated traffic being a function of the reduction in cost, or effective distance, dD.

Secular Growth

Long-term trends are notoriously difficult to project into the future, but as with all long-term planning some estimates have to be made. Two of the most important factors in determining traffic growth are population and income levels. If future trends in these are estimated and applied to a perfected formula similar to that given above a close approximation can be made.

CHAPTER SIX

THE LEVEL OF EXPENDITURE

In the last five chapters we have analysed the criteria on which highways should be planned to achieve an optimum in the sense that every project which is worth while is executed in that form which gives the greatest net benefit in excess of the costs involved. We must now examine how such projects are to be financed.

It is important to note that the anticipated revenue from a project in relation to the expenditure incurred on it is in no way a criterion of whether that project is worth while. Various forms of "solvency quotient" have been devised by writers in the United States,[1] and the idea is prevalent in the minds of many actually responsible for planning. It relates the anticipated earnings from a project, in the form of increased highway user tax revenues, with the expenditure on the project. There are two objections to use of this as a criterion of planning. In the first place, user taxes are not a measure of benefit. Straightening a curved road, for example, might reduce fuel consumption and fuel tax receipts, showing negative revenue from the project while the benefit might be considerable. Secondly, it assumes existing user taxes to be at the ideal level. If a project is solvent by this criterion it means that users are prepared to pay the expenditure incurred on the project, though no account is taken of costs which do not show themselves in direct expenditure, nor of benefits in excess of user taxes. If it is not solvent, however, it might mean that tax rates are so low that users will not be charged the full expenses involved, though they might be prepared to pay them. Alternatively, it might mean that tax rates are so high as to discourage traffic which would be prepared to pay the costs of the project, and would do so if rates were lower. Finally, it might mean that users are not prepared to pay the costs; in other words, the project is not worth while. Thus an adverse solvency quotient might not mean that the project is not worth while, while a favourable one does not prove that total benefits are greater than total costs; it only proves them to be greater than out of pocket expenses. It therefore approaches the problem in reverse by saying that a project must be solvent at existing rates of taxes in order to be worth while, rather than that a project which is worth while can be made solvent at the right taxes.

The method of determining whether a project is worth while was outlined in chapter IV. If it is worth while, it will be executed and the full costs must be met. The costs of highway transport fall into four groups: costs of the highway itself; costs of vehicle operation; personal costs of travellers; and community costs. The second and third categories will be borne by the users of the highway; the highway authority is concerned only with highway and community costs. All highway costs will enter into expenditures of the highway authority and are easily reckoned. But only some of the community costs will be in the form of direct expenditure by the highway authority; others will not. In this chapter we shall try to define the actual expenditure of the authority, and in the next chapter the distribution of this over time. Chapter VIII examines how the burden of this expenditure should be distributed between beneficiaries, and chapter IX deals with the means by which revenues can be collected.

The Aim of the Authority

Highway improvements inevitably involve gains to some people and losses to others. It has been argued above that these are all items of cost and benefit of the project and must be included at estimated values in the planning analysis. The question remains to what extent losses should be the object of financial compensation.

On simple grounds of equity it is reasonable to argue that persons worse off as a result of the project should receive full compensation. If the benefits are greater than the total costs the beneficiaries can meet all such claims and still be better off themselves. If benefits are less than costs the project will not be executed. Against this it can be argued that the free market takes no account of some costs and it will lead to an uneconomic use of resources if highway users are placed under an obligation to meet these, while competing forms of transport are not. This is really an argument against including in the planning analysis any costs which would not be considered relevant by private enterprise, and as such has been examined in the previous chapter. Such items are components of total costs and their inclusion in the analysis therefore results in a more reliable approximation to the optimum. That some lack of balance then exists between the public and private sectors of the economy is a good argument for some means to take account of these items in the private sector. As such it is irrelevant to this analysis.

The conclusion therefore is that full compensation should be paid to all persons in any way adversely affected by a highway project so as to leave them neither better nor worse off, to whatever extent this is practicable. The extent of this practicability is best determined by examining each item separately.

Compulsory Purchase of Right of Way

The cost of right of way used was discussed in chapter V, and the means of valuing it for planning purposes is an appropriate way to determine the amount of compensation. This is at the market value of the land without the highway improvement, plus disturbance costs. Two neighbouring sites might then be treated unequally in that one receives compensation so as to be neither better nor worse off as a result of the improvement, while the other derives increased value by reason of improved access. This benefit might be removed by recoupment or property taxation, which will be examined later, but there is no reason to pay more for the land actually acquired. Incidental benefits are bound to arise for some persons as a result of the improvement, and it will be impossible to avoid these or to give everyone else in the economy a similar benefit. The primary consideration is that as far as possible no one should be worse off.

There is a growing practice in the United States for the state highway departments to acquire land well in advance of requirements to avoid paying the increase in market value which occurs in anticipation of the improvement.[2] From 1952 to 1954 some $19,000,000 was spent by California on land which would have cost approximately $114,000,000 if not acquired until needed. The increase in market value of $95,000,000 was due almost entirely to betterment which the highways were expected to bring, and the Highway Department would have needed to pay the higher figure only because of the practice of paying, for land purchased, the current market value of adjacent land. It was shown in chapter v that the true cost of right of way is the market value of the land if no highway is to be built, and if compensation is assessed on this basis there is little if anything to gain by advance purchase. Advance purchase of neighbouring land to be sold later at the higher figure, generally called recoupment, is a method of taxing betterment which will be discussed below as a form of revenue.

Highway improvements can bring about reduced land values in other locations by diverting traffic and trade from them. Where a road is abandoned or closed there is a clear case for compensation for loss of access, but where access is not affected it is difficult to define the extent of compensation. In an extreme case the loss to an individual might be considerable. In one case of an airfield extension a section of main road was closed and a minor route improved round the enlarged airfield. As a result a café on the main road close to the airfield was left at the end of a *cul de sac* with no traffic and no trade; yet none of the café site was acquired and access to it was not affected. This loss of trade was part of the cost of the project and equity would demand compensation adequate to enable the café proprietor to move to a similar site on the new main road. There is no difference in principle, however, between this case and the case where a bypass relieves downtown congestion leading to an influx of business traffic and a fall in the trade of suburban businesses. Clearly it would be impracticable to assess compensation to each suburban business. Further, no account of the effect of such trade movements is taken by other sectors of the economy. The location of a bus stop outside one store can increase its trade at the expense of the competitor located between bus stops; if a railway station is closed the drug store near it might lose a part of its trade; yet no compensation is payable. In practice a distinction must be drawn between direct losses caused by the project for which compensation should be paid, and the indirect ramifications which are part of normal business risk. The only clear distinction is between cases where part of the site is acquired or access is affected, and those where they are not; but this is a purely arbitrary line for the sake of practical convenience and has no theoretical foundation.

Community Costs

The indirect effects of highways on the community as a whole were discussed in chapter v. Many of them are beneficial and their use as a source of revenue will be discussed later. The disadvantages of spoiling a beauty spot, the barrier caused by a limited access highway, noise and smell nuisance, and so on are costs which have been included in the planning analysis, but in so far as they are not reflected in other more direct costs—for example, a recognized beauty spot in right-of-way cost—it is not normally practicable to pay financial compensation. It is impossible to make a payment to every hiker through a valley because the view is less inspiring with a road in it. Normal rates of property tax on the right of way should be paid, however, to maintain consistency with other land uses.

We therefore conclude that the actual expenditures of the authority on the improvement should reflect the full costs as far as possible. They will include compensation for any land acquired and any loss of access, and the full construction, maintenance, and administration costs of the highway, together with property taxes on the land used. Many items of community cost will be excluded, however, because it is impossible to represent them by money transfers.

CHAPTER SEVEN

THE ANNUAL MONEY COST

In the last chapter we determined the expenditure actually made by the highway authority on a project. In order to pass benefits on to consumers in such a way as to lead to the optimum economic use of the facility, neither profit nor loss should be aimed at by the authority. As we shall see later, circumstances may arise in which losses on a certain project become unavoidable, or profits desirable to discourage submarginal traffic. Further, any pricing or taxation system will involve a certain amount of averaging, resulting in some projects making profits and some losses. But these are all the results of institutional imperfections; the aim should be the ideal position in which the beneficiaries from each project just pay its full costs, no more and no less.

In the planning analysis the cost stream and benefit stream over the anticipated life of the highway were discounted to current values. In practice the expenditure and benefit streams will be widely divergent. Generally speaking, the benefits will increase with traffic growth, while the expenditures will begin with the initial high capital outlay, followed by a period of low annual costs, rising later as maintenance costs increase on an aging facility. A given year's expenditure cannot therefore always be attributable to that year's traffic, and some means of reconciling the two streams must be devised. The real question is how much revenue should be collected each year.

The expenditure can be broken down into that which varies with the volume of traffic and that which does not.[1] The variable costs are directly related to the volume of traffic, and it is clearly appropriate that each year's traffic should bear its variable costs. The remaining costs are a function of time and the type and capacity of the facility. Since type and capacity are given, as determined by the planning analysis, these costs become a function of time. The time cost is the annual amount required to amortize fixed expenditure on each item over its anticipated life. Right-of-way and initial development costs will never need replacing and their life is essentially the anticipated time the highway will remain in use. In practice highway routes are rarely abandoned, but changing demand patterns can considerably affect useful life. The period should therefore be limited to the forseeable future in which there is no reason to fear that they will become obsolete. It is prudent to err on the side of underestimating this so as to avoid placing a debt burden on the distant future which might be unable to meet it. The maximum period should be in the order of twenty to twenty-five years, but on minor roads or roads in areas with uncertain economic futures considerably shorter periods might be appropriate. Construction and maintenance costs which are not variable should be amortized over the life of the structure, but in no case may this be longer than the anticipated life of the route. Administration and property taxation will be annual items.

The fixed or time costs are essentially the costs of having the highway available for a year irrespective of the amount of traffic which uses it. There is normally therefore no justification for extracting higher contributions from the years with more traffic; each year's cost over the amortization period is equal and the revenue to meet it should be equal. There is one circumstance, however, which might justify lower rates in the earlier years. This is where a large part of the anticipated traffic will be generated by economic development stimulated by the highway project, but materializing only after a time lag. In the first year costs might be greater than benefits, but the effect of one year's running at a loss is increased traffic in the future. Revenue can never be greater than benefit and could not, therefore, equal annual cost for the first year. This case must be distinguished from that where traffic growth is expected to justify the project, but where the growth is not dependent on the existence of the highway, that is, where the project is justified by secular growth rather than generated traffic. In this case if the first year's benefits are below costs it will be better to postpone the project. Thus, there is justification for making payments higher in the later years of the amortization period when the project is a genuine "infant industry" case. There are serious dangers in adopting this practice where it is not appropriate, however. Where appropriate it is prudent to make the earlier contributions as high as is consistent with not discouraging optimum use of the facility. In all cases each year's traffic must bear its variable costs.

Historic Costs

It is often argued that historic costs are irrelevant, but it is somtimes ambiguous what they are irrelevant to. It is therefore worth examining what they are and to what extent and in what circumstances they are important. Historic costs are past, once-for-all, irredeemable overheads, the benefit from which is still being reaped, but where the assets have no, or very little resale value. Once a cutting is cut the cost is incurred and the expenditure cannot be recouped. Very little, if any, continuing expenditure remains. Since the cost of the cutting is in no way affected by whether it is used, should the users continue to pay for it?

From the planning standpoint irredeemable historic costs are totally irrelevant once made. The only consideration relevant to the question whether to abandon an existing right of way in favour of another is whether the resulting difference in benefits is worth the resulting difference in costs. The cost of the cutting on the old route will not be affected by whether it is abandoned, and will not therefore be an item of difference. The cost of any new cutting which will be needed on the new route is an item of difference, however, and is important until actually made.

From the financial standpoint, however, the position is somewhat different. Once-for-all costs were incurred only because it was believed that benefits resulting from the project would be adequate to amortize its cost over its useful life. If this assumption proves correct there is no reason why the beneficiaries should not pay the cost. Indeed, if they do not it will be extremely difficult to raise funds for similar projects in future when it is realized that they will not be repaid. Someone has to pay the cost of fixed capital investments and where they result in benefits greater than their costs, justice demands that the beneficiaries meet the costs. This principle may lead to some complications, however, as in the following examples.

Benefits on an existing road are adequate to pay variable costs and amortization of the historic costs of a cutting, but the additional benefits of a proposed re-routing will be more than the additional costs, so the old route is abandoned. If annual variable costs of the old route are v and amortization of historic costs, h, benefit, b, is equal to or greater than $v + h$. Annual total costs of the new road, t, minus savings in variable costs on the old road, v, are less than the increase in benefit, b'. That is:

$$b > v + h$$
$$b' > t - v$$
$$b + b' > t + h$$

Therefore benefits from the new road are adequate to pay full costs of the new road and amortization of the historic costs of the old road. Should payments still be made for the amortization of the historic costs of the cutting on the old road when it is no longer in use? Clearly, the fact that the old route has been abandoned means that it is obsolete, i.e., that although the cutting is already there it is not worth using. It is therefore yielding no benefit; the anticipation that its life would be long enough for its initial costs to be fully amortized at the rate so far charged has proved wrong; and there is no reason why persons who have discarded the cutting as of no further use to them should continue to pay for it.

An intermediate case arises where the difference in benefit is less than the difference in cost which would result from construction of a new road, and the old route is therefore retained. But the new route would yield a greater net benefit after paying total costs than the old route yields after paying variable plus historic costs. This is the case where if the cutting had not been cut it would not now be worth cutting, but since it is there it is worth using. It appears that since the users of the cutting do not consider it worth its full amortization costs, because they can now get greater net benefit from a different route, they should not pay them. But since the cutting still yields them some benefit they should pay what it is worth, that is, what it would cost to get that benefit by other means. If we let h' be the amount they should now pay it can be shown that, given $b' < t - v$, and $b + b' - t > b - v - h$, then the old route will yield the same net benefit as the new route would where $b - v - h' = b + b' - t$; that is where $h' = t - v - b'$ or where the amount actually paid, h', toward the amortization of historic costs is equal to the amount by which the additional costs, $t - v$, of a new road would exceed the additional benefits, b', which it would yield. This principle is consistent with the previous example, where $b' > t - v$, since in this case the additional costs would not exceed the additional benefits and the contribution towards amortization of historic costs on the old road would be zero. In a case where $b + b' - t < b - v - h$, i.e., where the old road yields benefit greater than variable and historic costs, and where no alternative highway would yield greater net benefit, the full amortization of historic costs would be paid.

Thus we conclude that future traffic should pay the full annual cost of past investments while they continue to yield benefits greater than these costs. Where they yield benefits less than costs future traffic should pay only what they are still worth; and where they no longer yield any benefits no payments should be made. Variable costs must always be paid, however, and there might be a certain element of variable or avoidable cost associated with historic costs. The cost of converting agricultural land to a highway is high, but once it is done the cost of converting it back would be high, as the surface and structures would have to be removed and the soil fertilized. Where the value in another use would be enough to pay these conversion costs and still have a margin of benefit, the opportunity cost of keeping the land in highway use is that net benefit foregone. Where the benefits of the highway are not adequate to cover this opportunity cost the right of way should be abandoned to another use. This item will usually be very low by comparison with the amortization payments on the initial investment. It remains an important concept, however, even when the initial outlay is fully amortized and the right of way completely paid for. There will then be no need for any financial payments, but just as proposed improvements were analysed in terms of benefits and opportunity costs, so will proposed abandonments. Where the anticipated net benefits of the highway in future are less than the net benefit which the land could yield in any other use, the highway should be closed and the land sold.

In those cases where benefits are inadequate to pay full amortization of historic costs the question remains who meets the losses on the investment which has proved a financial failure? This will largely depend on the source of the initial capital and the rate of interest paid on it. The capital might have been privately subscribed by a bond issue backed only by revenue from the project, as in the case of some toll roads and bridges, or backed by taxes on the highway user, or by the general exchequer. In the first case the rate of interest would reflect the risk involved, and if toll receipts are inadequate to pay amortization the loss by the investors can be regarded as a normal business risk. In the second case the burden would fall on other highway users and in the third case on the general exchequer. If the initial capital were raised from a road fund tied to highway user tax receipts the burden would fall on other highway users; while if it were raised from the general exchequer it would fall on the general taxpayer. In all cases we can say that where the rate of interest used takes full account of the risk involved, the investor should stand the loss. Where the repayments are backed from more general funds so as to secure a lower rate of interest the burden can be arranged to fall either on the general taxpayer or on other highway users. By pledging all highway taxes for loan repayments, highway users are pooling the risks on a simple insurance principle, and must jointly stand financial responsibility for any failure. In no case can the general taxpayer be justly held liable for the losses. It might be argued that putting the burden onto other highway users involves extracting a profit from some roads to pay losses on another—a cross-subsidy—but this is not really the case. The high interest rates resulting from risk are part of costs, not profit. Where an over-all saving results from the pooling of risks, the burden on other highway users of paying for failures on some projects will be offset by the lower interest cost of capital for other highways. The benefit of keeping interest rates lower than the risk element would necessitate, by pooling risks, goes entirely to the beneficiaries of highway projects. The cost of meeting an occasional bad debt must likewise be theirs; it cannot be justly shifted to general taxpayers without involving a clear case of subsidy to highways in general. Whether it is better to borrow money at a high rate and let the investor take the risk, or at a low rate and pool the risk between projects will depend on whether the market rate reflects a state of undue optimism or undue pessimism in the prospects of such projects. If the full risk rate is paid it will be higher the longer the period of amortization, since the risk of default increases over time, more distant forecasts being less reliable. If the lower rate is paid, a long amortization period will increase the probability of the project becoming obsolete before recouping its costs, that is, of the burden being shifted to all highways users. In either case therefore it is prudent to amortize long-term investments as quickly as is compatible with not discouraging marginal traffic by excessively high taxes in the early years.

Sources of Capital

Highway costs can be divided into regular annual costs and the initial long-term investment. The latter is amortized over the assumed life of the asset and must therefore be paid for before it is recouped from revenues. There are three sources of capital for such investments: a highway fund tied to highway user tax receipts, the use of which is normally called "pay-as-you-go" financing; bond issues on the open market; and the use of funds in the general exchequer. Where the latter is meant as a direct subsidy it is pay-as-you-go financing, where it is to be repaid to the exchequer from highways taxes it is loan financing. What then are the relative merits of these two systems?

Some of the arguments often raised in this discussion are irrelevant to the real question and can be quickly dismissed. The dangers of using too long a repayment period, thereby burdening the future with an inflexible debt structure on projects which might become obsolete, have been discussed above. It is also argued that bond issues provide capital so readily that overinvestment in highways might result. This will be guarded against by use of the planning analysis outlined in chapters II to V. The fact that loans are justified only by estimates of the future which might prove inaccurate does not make loan financing as such unsound, since the same is true of any form of financing. It is important to distinguish the justification of an expenditure from justification of the use of a particular source of capital. The difficulty of making long-term forecasts for a project is not affected by the type of financing used. The argument that committing future funds will prevent diversion or dispersion of highway revenues does not apply if a sound administrative approach is applied to these problems anyway. The interest cost of loan funds is not an additional cost of bond financing. There is an interest cost involved in using any capital which has alternative uses whoever owns it, and only projects that show benefits adequate to meet interest costs should be undertaken. We can therefore limit the discussion to those cases where prudent use of loan funds could be made. This will be the case if existing reserves in the highway fund are inadequate to undertake all worth-while projects. It is worth noting in passing that these reserves can have been built up only by charging historic costs on past investments either where these were financed from the general exchequer, or where they had already been fully amortized. The highway fund is an accumulated stock of capital which can be used to finance new projects, the returns from which finance future projects, and so on indefinitely. Proposals to use loan financing arise only where the fund is inadequate. It may be perpetually inadequate over a period of expansion, or have recurring deficiencies where the highways built in a past period of rapid expansion all fall due for replacement at the same time. The problem will be most pressing when a backlog of work has built up. This was the position in 1945 following a period of minimum road expenditure during the war years, which in turn had inherited a backlog from the depression years of the 1930's. The postwar backlog coincided with rapid developments in the use of motor vehicles and a generally expanding economy. The field for sound investment in highways far exceeded the available capital in highway funds, while rising costs depleted the value of any reserves which had been built up. Three courses of action were then possible. Increased highway user taxes could be levied to raise capital quickly for urgent projects in accordance with a pay-as-you-go policy; the capital could be borrowed and repaid from user taxes over a number of years; or roads could not be built. Much controversy in the United States split the various states between the first and second policies, while the United Kingdom adopted the third.

The disadvantages of a period of high user taxes to pay for the high initial investment in highways are both practical and theoretical. This method leads to frequent changes in tax rates, which do not reflect changes in annual costs, and causes considerable inconvenience to the motor transport industry. This problem has not been apparent in California, one of the strongest exponents of this policy, because high tax rates have been maintained to finance a continuing high level of investment in highways. Theoretically "pay-as-you-go" is really "pay-before-you-go", and involves a gift by current highway users to the future. Even those who will still be highway users to recoup the benefit are subjected to forced saving, whereas a bond issue is essentially voluntary saving. Where the money is borrowed from the general exchequer it is the general taxpayer who is forced to save. Where the whole highway system is taxed to pay for one new project it involves a gift by general highway users to the future users of that project. It is impossible to tax the users of the project in question before it is built. Perhaps the greatest disadvantage with pay-as-you-go financing is its inability to deal with a really major project. The Ohio Turnpike was completed in less than three years at a cost of over $280,000,000. Annual State highway expenditure was in the order of $28,000,000. To raise the full cost of the turnpike over the period of its construction, without detracting from other projects, would have called for more than a fourfold increase in highway user taxes for three years. This would clearly have caused considerable problems for the road transport industry—not to mention the political repercussions.

The disadvantages of deferring projects until the necessary capital can be accumulated from excess taxation are obvious. The loss of benefit from the road during the period of waiting can be considerable, and if the project is a sound one the increased benefits to traffic would have more than paid the interest costs involved in borrowing the capital. Delaying replacement of a highway when its economic life has expired leads to unnecessarily high maintenance costs. The continuing shortage of capital leads to the adoption of plans with low initial costs even though high future maintenance costs will be involved. Piecemeal construction of a new network as funds become available is less satisfactory and more expensive than a single large project, and land values rise before acquisition with anticipation of improvements.

The prudent use of loan financing can overcome these difficulties by allowing the immediate construction of expensive projects as soon as they are worth while, concentrating investment in periods when interest rates are low. Furthermore, the optimum plan from a long-term standpoint can be adopted, which in the case of modern arterial roads means heavy capital overheads and low future maintenance costs. Obsolete plant can be scrapped as soon as replacement is more economic than continued maintenance. The burden of expenditure is more equitably distributed, being spread over the life of the asset so that those benefitting from the project in future bear the cost of it. It also facilitates the influx of capital to highway construction from other sectors of the economy, or from other countries. This might be very important for the underdeveloped countries and for development areas. It has even been proposed that capital for the modernization of English roads might be borrowed in the United States.[2]

Loan financing is not without its difficulties, however, some of which inevitably go with its advantages. By enabling a short period of rapid construction at a time of general economic expansion it can cause a boom in the construction industry with attendant rising costs of labour and materials and falling off in competition for tenders. This might be followed by a period of lesser activity causing a slump in the construction industry. By comparison, pay-as-you-go financing leads to a more stable level of expenditure spread evenly over time. A further advantage of undertaking a major modernization programme in stages is that it remains flexible for later stages to be modified in the light of experience. This argument is often used by politicians as an excuse for inactivity, however, when it really resorts to the fact that if nothing is done no mistakes will be made.

Thus, we can conclude that long-term highway projects are as suitable a field for long-term bond financing as any other form of long-term industrial investment. The use of this method can be of great value in periods of expansion, and is combined with the economic advantage of coincidence of benefits and costs. There are, however, severe dangers of misuse of the system, and these must be guarded against. The project must be sound, and reasonably certain to be able to amortize its costs over a period during which the benefits from it can be confidently predicted. Excessive use of bond-financed construction over short periods of time can lead to cyclical instability in the construction industry, which will perpetuate itself since the projects of one period of intense activity will fall due for replacement about the same time. The rate of expenditure on worth-while projects should therefore be limited not by the availability of capital, since loan capital is readily available, but by the capacity of the construction industry.

CHAPTER EIGHT

DISTRIBUTION OF THE COST BURDEN

IN THE PREVIOUS TWO CHAPTERS we have discussed the items of actual expenditure incurred in highway construction, and the distribution of these over time. Given the annual money cost to be recouped we now have to examine the distribution of the burden among the beneficiaries of the project. We are not at this stage concerned with the functions and responsibilities of different levels of government; it is assumed for simplicity that there is only one highway authority.

Three principles can be derived from the various methods which have been proposed for allocating the cost burden. The first is an equitable distribution aimed to spread the burden fairly between those who benefit. The second is to approximate the charging system to the methods normally adopted by a free market; this means that only the direct beneficiaries from a project contribute to its cost. Finally, the distribution should be economically expedient by not causing undesirable incentive and disincentive effects which would upset the normal balance of economic forces. In practice there is often a fourth factor of political expediency which is sometimes dominant, but this is entirely devoid of economic justification and totally irrelevant to this analysis.

The allocation of total costs is normally undertaken in two stages, dividing them first between the three sectors of the economy which benefit; vehicle users; property owners; and the community in general; and, second, between the individuals and groups in each sector.

The so-called community benefits which are often held to justify contributions to highway costs from general taxation expose some confusion of thought. Postal services, school busses, police and fire protection services all involve use of highways by public vehicles; but the only sense in which this is a community benefit is that the vehicles are publicly owned; analytically they are vehicle benefits.[1] Similarly, the increased economic prosperity and improved standard of living brought by good highways result from the benefits to traffic; they are not a separate item justifying general subsidy. The valid community benefits can be subdivided into those of general applicability to the whole nation, such as unemployment relief and defence, and those of purely local benefit, like sidewalks, sewers, and so on. Some items are difficult to classify as general or local—slum clearance schemes, for example—and it will be entirely a political decision to what extent these are centrally or locally financed. The contributions to highway costs in respect of general benefits will come from general taxation, and of local benefits from local taxation.

Benefits to property result from improved access and reduced transport costs of goods and persons to and from the site. As we saw in chapter v these arise entirely from a shift in the incidence of vehicle benefits, and though it might be convenient to tax some vehicle benefits at this stage, benefits to property are not a distinct form of benefit.

Vehicle benefits result from reduced operating costs and can be further subdivided into benefits to the various classes of vehicles and the individual vehicle users within each class.

Four general principles of taxation are used in the literature of highway finance[2] and are reflected in various highway taxes. They are: taxation on the general basis of ability to pay; taxation based on the value of service provided; taxation according to the amount of use made of the facility; and taxation based on the costs incurred by the highway authority in providing services. Each of these deserves examination in relation to the division of costs both between sectors of the economy and between individuals within each sector.

Division of the Burden between the Community, Property Owners, and Motor Vehicle Users

The Ability-to-Pay Principle

The ability-to-pay principle has sometimes been confused with the benefit principle by defining a person's ability to pay as the net benefit he derives from the project. As an over-all principle of taxation, however, ability to pay is not related to any benefit directly received from highways, any more than it is directly related to the benefits from other forms of government expenditure. It therefore regards highways as a general government service with the cost falling on the community as a whole irrespective of their use of highways. This approach was examined in chapter I and discarded as unrealistic under current conditions.

The objections to use of ability to pay as a basis for highway taxation are both economic and political. No direct charge would be made for use of highways, which means that journeys would be undertaken the value of which is only just sufficient to cover operating costs. The marginal cost of highway use, both in damage to highways and losses to other users resulting from increased congestion, are not considered and the total cost of these trips may well be greater than their value. The absence of a direct charge for highway use therefore leads to uneconomic use of roads by submarginal traffic.

Any use of ability to pay as a taxation base involves, on the whole, a subsidy by the rich to the poor. This might or might not be desirable on social welfare grounds, but in any case it should be kept separate from pricing policies for public utilities. This is particularly important in a field such as transport where a subsidy to highway users would upset the balance of competition between highways and other transport media. Sumptuary taxation and subsidies to highways are discussed in chapter X below. Ability to pay might be a valid principle for allocating the community's share of costs among individuals, because the very nature of community services is that they are spread evenly or at random throughout the community. The normal principles of sumptuary taxation are therefore applicable in this instance. But it gives no guide at all to the proportion of highway costs which should be borne by the community, property owners, and motor vehicle users respectively; and is not an appropriate way to subdivide the property and vehicle shares among individuals.

The Benefit Principle

Any direct highway tax must be a tax on benefit in the sense that if benefit were not adequate to pay the tax the trip would not be undertaken. If the primary aim is achieved that no one should be worse off as a result of the project, then all taxation for the project should be covered by benefits received from it. The benefit principle as normally propounded, however, goes further than this, and maintains that charges should be proportional to benefits received. Since benefits exceed costs on any worth while project this proportion would be less than one.

As a method of allocating costs between the three sectors of the economy the benefit principle suffers from serious difficulties. All benefits to property and some community benefits are shifted from motor vehicle benefits. It is difficult to assess the amount of this benefit which is shifted in any particular case, and the proportion will be affected by the taxation system used, the benefit tending to move to the sector taxed. This can be seen by taking a simple example where an expressway is expected to open up a new district for residential development. If we assume the district to contain homogeneous sites, each adequate for one family containing one commuter, all commuters making the same daily trip and therefore deriving the same benefit from the highway, the market for sites in terms of annual rent can be shown as in diagram 15.

Diagram 15

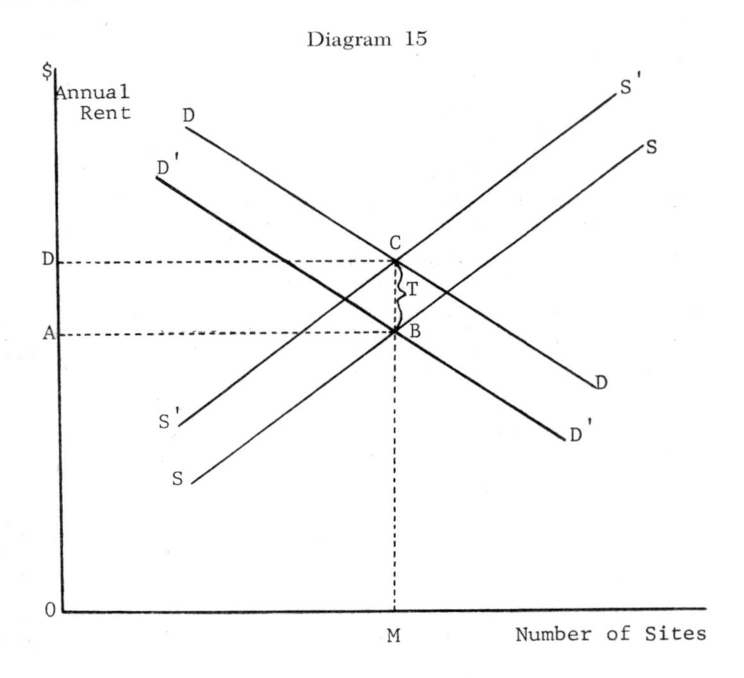

Lines *SS* and *DD* are the supply[3] and demand curves for building sites after construction of the highway but before either vehicle or property taxes are levied. If vehicle taxes are imposed such that each family pays $\$T$ per annum from the net benefit it derives from the road, the demand curve for sites in the district will fall by T throughout its length to $D'D'$. If no vehicle taxes are imposed so that DD remains the demand curve, but property taxes of $\$T$ per annum per site are levied on landlords the supply curve will rise vertically by T to $S'S'$. In either case OM sites will be occupied and $ABCD$ taxes paid. If vehicle taxes are used rent will be OA; if property taxes are used rent will be OD which is above OA by the amount of the tax. Thus the amount of benefit T which is to be collected in taxes will shift to whichever sector the taxes are levied on. This same tendency will operate in a practical case of heterogeneous sites and families. Any benefit to property results from a shifting in the incidence of vehicle benefits. It is therefore the same source of benefit which is taxed whether the tax is levied on vehicles or property, and the distribution between vehicles and property of net benefit after tax will be the same no matter how the tax is divided between them. The benefit principle can therefore give no guide to the distribution of the tax burden between vehicles and property. It remains to see whether the benefit principle can determine the share which should be borne by the community in respect of those benefits which are not shifted from vehicle benefits.

Great difficulties would arise in identifying and evaluating all the items of community benefit, but assuming this could be done there would remain two objections to use of the benefit principle. Some of the community benefits will be the removal of "worsement" originally brought by motor vehicles. The increase in motor vehicle traffic may have brought considerable dust nuisance to neighbouring property on unsurfaced roads. A dust-free surface on these roads would bring community benefit by the removal of the dust nuisance, but unless compensation was paid when the dust nuisance arose, it appears inequitable to charge for its removal.

A more important theoretical objection arises where a contribution from the community for community benefits would upset the balance of economic forces between industries. We included the value of highways for defence purposes and unemployment relief as community benefits, and if their inclusion results in the adoption of a plan which would not otherwise have been optimum, the community should bear the difference in costs, just as it subsidizes uneconomic agriculture. But to a large extent plans which would have been optimum anyway will still have community benefits for defence, employment, and the like. This is not peculiar to highways. The steel industry, shipping industry, oil refineries, railways—all are of community value for defense purposes, while telephones are of great value to fire protection and police services. No subsidy is paid to these industries to represent community services, however, and economic competition between industries for resources will be upset if highways are singled out for such payments, as they would be under the benefit principle.

Thus we conclude that the benefit principle is of no value in allocating a share of highway costs to property, nor to the community benefits which result from vehicle use. If used to determine the share allocated to the community for other benefits it will result in a subsidy to highways which no other industry enjoys and thereby upset the balance of economic forces.

Amount-of-Use Principle

There are three bases for distributing highway costs according to use: relative use, predominant use, and availability for use.

The relative use principle relies on traffic surveys to determine the proportions of local and through traffic on any road. The through traffic proportion is paid by motor vehicles, and the local proportion by the community. Roads are normally classified into groups with different proportions of vehicle and community contributions for each class. No distinction is made between community and property benefits, but the community's contribution is normally paid from a property tax. This system works in practice, but that does not make it economically sound.

The major objections to this principle are to the assumptions on which it is based. The distinction between local and through traffic and the arrangement of classes are entirely arbitrary and are governed by what is practically convenient rather than what is theoretically defensible. The traffic survey does not separate vehicle from community or property benefits, but local from through traffic. It is assumed that the share of highway costs attributable to local traffic should be borne by the community, and that the incidence of the benefits to local traffic is the same as that of local taxation. At best, therefore, it is a method of dividing the motor vehicle share of costs between local and through traffic. It is a rash assumption that all local, and no through, traffic shifts benefit to property owners and the community in general or that any errors are necessarily self-cancelling, and the aggregate amount attributed to the community may therefore be wrong. The division of the community's share, once this is determined, between property owners on the assumption that the benefits from local traffic are divided in proportion to ratable value, while purely arbitrary, is probably as reasonable a basis as any, however. As we have seen, where benefits are shifted they tend to shift to the sector on which taxes are imposed, and it does not matter in aggregate terms therefore where the benefits to local traffic are taxed. The only community benefits worth distinguishing are those which are not shifted from vehicle benefits, and the relative use principle takes no account of these.

The predominant use principle is based on the same assumptions as the relative use principle, but is simplified by having only two classes. Highways which carry a majority of through traffic are paid for entirely by motor vehicle users, and others entirely by the community. It suffers from all the faults of the relative use theory plus the unjustified assumption that the local share of predominantly through traffic routes will equal the through traffic share of predominantly local routes.

The principle which bases taxation for highways on availability for use is inappropriate to a division of costs between sectors, since highways are always available for use by all sectors. This could be held to justify charging each sector equally or each person in each sector equally, but neither would take account of either costs or benefits.

The Incremental Cost Principle

The essence of this principle is to take a basic standard of highways the costs of which are attributable to one group, and assess the extra costs of higher standards as attributable to the groups benefitting from them. To divide costs between motor vehicles and the community, one proposed method is to attribute the level of costs before the motor vehicle age to the community and the remainder of present costs to motor vehicles. There are practical difficulties in determining a suitable base year and standard, and in adjusting for changes in the value of money, but it is on the unrealistic assumptions and theoretical objections that this principle mainly falls down.

It assumes that the standard which was adequate for the community before the motor vehicle age would be adequate today if motor vehicles had never been invented. Many different speculations can be made as to what would have happened, but there is no evidence to support the assumption that standards would have remained constant. Even if standards would not have improved in the absence of motor vehicles, the community still derives some benefit from the improvements which have taken place, while the cost of them would be borne entirely by motor vehicle users. It is further assumed that the community was entirely responsible for costs before the advent of motor vehicles, while in fact horse-drawn traffic cannot justly be absolved from all responsibility for all highway costs in those days. The extent of this responsibility has been taken over by motor vehicles. In cases where highways have not improved despite the growth in motor vehicle traffic, motorists would pay no share of the costs although deriving considerable benefit from use of the roads. Where standards have improved the motorist bears the entire extra cost, so that the burden of any extra costs which have been incurred because of the community benefits they will yield, will fall not on the community but on motor vehicle users.

The assumptions are so unrealistic that this application of the principle cannot be considered valid. The assumption that differences would cancel out in practice is as administratively convenient and as analytically inaccurate as the assumption that the historic costs of the highway system which the motorist inherited from the community just balances the current responsibility of the community for highway costs.[4]

The objections, however, are to the generally advocated misuse of the incremental cost principle in this way, not to the principle itself. The misuse results from attempting to allocate joint costs by a method which applies only to specific costs. Instead of arguing that the community should pay what highways would cost them in the absence of motor vehicles and motor vehicles the rest, we could argue that motor vehicles should pay what it would cost them for a road system not used by the community and the community the rest. The incremental half of each of these arguments is valid but the allocation of joint costs entirely to the other group is not.

Highway costs can therefore be divided into three categories: those incurred entirely for the benefit of motor vehicles which should be borne by motor vehicle users; those incurred entirely for the benefit of the community, which should be borne by the community; and those which are joint costs and yield benefits to both. The incremental cost principle applies to the first two types and is valid. It does not apply to the joint costs and gives no guidance as to how this group, which is normally the largest, should be divided.

Distribution of the Motor Vehicle Share of Costs between Individual Users

The Ability-to-Pay Principle

This was discarded as an inappropriate way to allocate costs between sectors, and is equally invalid for the allocation of costs between individual motor vehicle users, for the same reasons.

The Benefit Principle

As we have already seen all vehicle taxes must be on benefit received and are limited to that benefit derived from highways over and above that from the closest form of competitive transport. Higher tax rates would only discourage use of the roads. Objections to attempts to make taxation proportional to benefit are both administrative and theoretical.

It is administratively impossible to evaluate the net benefit derived by each vehicle user from the highways, since so many factors affect the value of service. Use of group data would inevitably lead to the anomaly that there would be greater differences within, than between, groups.

The great theoretical objection is the same as that to the ability-to-pay principle, that it encourages the uneconomic use of resources. Trips which yield benefits only just enough to justify operating costs would yield no net benefit and would therefore make no contribution to costs by this principle, while the highway and congestion costs caused might be considerable. Any principle which does not include direct costs as one basis of taxation will suffer from this fault.

Apart from these difficulties the method would be politically unacceptable as it is essentially a form of charging what the traffic will bear. Economically there is no objection to price discrimination in some cases, but where this takes the form of atttempting to charge the man driving his sick child to the doctor more for use of the road than the pleasure tripper, simply because the journey is worth more to him, there may well be political difficulties. Since the costs of both trips are equal as far as the highway authority is concerned it would seem inequitable to charge differently. The contradiction between the benefit theory and public opinion in such cases is well illustrated by the reaction to an increase in the fuel tax. It is generally felt that it matters less about luxury pleasure travel being highly taxed than essential business traffic. The luxury traffic, however, is presumably that with little net benefit to cover the tax, since it is supposed that it could cease to flow without much loss of benefit; while the essential traffic is that with a substantial margin of net benefit out of which to pay taxes, since its stoppage would cause serious loss. As long as highways are tied up with political responsibility, it is important not merely that economic justice be done but that it be believed to be done. The benefit principle would achieve neither aim and we can therefore discard it as an inappropriate basis on which to distribute highway costs.

Amount-of-Use Principle

The idea that the more a person uses highways the more he should pay towards their costs appears at first sight so equitable that its difficulties are often overlooked. It gives rise to both practical and theoretical problems, however.

The first difficulty is to measure use. The gross ton-mile and the passenger-mile measure neither the value nor the cost of highway service, and no satisfactory scale exists of how many passenger miles there are in a ton mile. Costs vary with weight, number of axles, number of wheels, speed, and so on, but they are not proportional relationships, whereas cost is directly proportional to distance. The damage inflicted by a 40,000-lb. truck covering one mile is more than that inflicted by ten 4,000-lb. cars. Gross tons squared times distance would perhaps be more accurate than the straightforward ton mile, but no simple measure of use can accurately reflect costs because it must ignore some important variables.

It would be impossible to keep a record of each vehicle without excessive administrative and enforcement costs, and grouped data would therefore have to be used, charging each vehicle the average for its group. In practice use by vehicles in the same group might vary more than use by vehicles in different groups. Grouping would in any case be an arbitrary process.

It has been suggested that operating cost is a useful measure of highway use, and a partial application of this is seen in the fuel tax. There are two major objections to this measure, however. By assessing contributions in proportion to operating costs it charges the heavy vehicle less per gross ton-mile.[5] A typical fouraxle, tractor-trailer combination, weighing about thirty tons loaded, covers about 5 m.p.g. or 150 gross ton-miles per gallon. A passenger car, weighing two tons loaded, covers about 15 m.p.g.[6] and only 30 gross ton-miles per gallon. The highway costs incurred by a vehicle are a function of impact force which in turn is a function of wheel load and speed. Wheel loads vary considerably with the design of commercial vehicles, but are invariably higher than for passenger cars, while speeds are generally slower. Most authorities agree that on the whole heavier vehicles incur higher costs on the highways than light vehicles per gross ton mile,[7] and they are certainly responsible for higher highway costs per gallon of fuel consumed than are light vehicles. The fuel tax therefore discriminates in favour of heavy vehicles to an extent not justified by the relative costs they incur on the highway authority.

Secondly, highway costs and vehicle operating costs are inversely related in that the object of spending more on the highway is to reduce operating costs. If different total amounts are spent on two otherwise similar highways with similar traffic, vehicles using that on which most is spent will have lower operating costs, but be responsible for higher highway costs. Operating cost is in this respect an inappropriate basis for assessing responsibility for total highway costs. It is important to distinguish construction cost from maintenance costs, however, since a road with high initial costs will suffer less damage from given traffic than a road with low initial costs. The marginal costs incurred on the authority directly by highway use will therefore be more on poor roads where operating costs are higher but where fixed highway costs are lower. Operating cost, therefore, varies in the right direction, though not necessarily to the right extent, as a measure of responsibility for marginal costs on highways of different standards, but in the wrong direction as a measure of average fixed or average total costs.

There remains one fundamental theoretical objection to taxation according to use. In a pure form it is average total cost pricing. If marginal cost is below average total cost it will discourage marginal traffic, while if the marginal cost is above average total cost (as it might well be under congested conditions) it encourages submarginal traffic, i.e., that which is not prepared to pay all the costs for which it is directly responsible.

Availability for use has been suggested as a method of allocating overhead costs and there is much to be said for the argument that since overhead costs by definition do not vary with the amount of use made of roads, users should contribute to them equally irrespective of the amount they use them, though contributions from vehicles of different weights might vary. This argument is often accompanied by the assumption that nearly all highway costs are overheads. When the costs to traffic caused by congestion are included however, this is far from true, and this principle can therefore allocate only part of highway costs.

The Incremental Cost Principle

We concluded above that the incremental cost principle was valid within its limitations, but that it could not allocate joint costs. The same is true for the allocation of the motor vehicles share of costs among users. Where costs are incurred to provide a service to one group of vehicles only, it is right and proper that that group should pay them. But it must be certain that the service applies only to that group.

This method is often advocated for the division of the cost of surfacing a highway, assuming that surface cost is a function of gross weight. In fact, it is a function of impact force which in turn is a function of weight, size, number of axles, number of wheels, type of roads, type of tyres and springs, speed, and many other factors. Of all single factors wheel load is perhaps the most important. The use of a highway by vehicles inflicting different impact forces can affect initial construction costs of surface width and thickness, and maintenance costs.

The minimum thickness of a concrete road surface is a function of the square root of the maximum wheel load using it.[8] Passenger cars require a thickness of 4 inches at the centre and 5.66 inches at the edges, and the heaviest commercial vehicles 6.31 inches at the centre and 8.92 inches at the edges.[9] The common fallacy is to argue that where the road is used only by passenger cars and the heaviest trucks, the trucks should pay the difference between the costs of these thicknesses. This, however, relies on the assumption that the minimum thickness would be used for passenger cars alone, which is often not the case. Initial construction costs and maintenance costs are inversely related; the more expensive the surface the less maintenance it will need. On a given surface, maintenance costs are largely a function of the volume of traffic. Thus where the volume of passenger car traffic is high it might be economical to use more than the minimum surface thickness to reduce future maintenance. Most of the New York State parkways, which are limited to passenger car traffic, are of construction standards suitable for mixed traffic including heavy trucks. In some cases, however, highways built for heavy vehicles can cost over 50 per cent more than if built for passenger cars alone.[10]

In some cases therefore the thick surface will be economical for passenger cars alone, and will not be incremental to trucks. It will be difficult in the case of any particular road carrying mixed traffic to determine what parts of the costs are really incremental and what are joint costs for all traffic. Other relevant factors also complicate the issue. The stronger the surface the lower the fixed depreciation from weathering, for example, and a strong surface will reduce variable maintenance costs even though it would not have been an economic proposition for passenger cars alone. Such reductions in costs for other groups must be deducted from any increased construction cost before it is allocated incrementally. Finally, it must be remembered that even where part of the surface cost can be allocated incrementally, other costs, with the possible exception of some structures, are not affected by surface thickness and remain joint costs.

The width of highways may be affected by traffic requirements in two ways; width of lanes and number of lanes. Both affect the total width and therefore the costs of grading and structures as well as surface costs. Wide vehicles require wide lanes, and heavy vehicles tend to be wider than passenger cars. But before the extra cost of lane width over the minimum necessary for passenger cars is allocated incrementally, allowance must be made for the fact that fast traffic also calls for wider lanes, and passenger cars tend to travel faster than heavy commercial vehicles. Thus the extra lane width might be as important for the fast as for the heavy vehicles and therefore be impossible to allocate incrementally. Even where the number of fast cars would not alone justify the cost of wider lanes, they will still benefit if these are provided for wider vehicles and should contribute something to costs.

The number of lanes required will be a function of the peak volume of traffic. This is incremental in a marginal sense, but it is impossible to identify the marginal vehicle. It is as fallacious to argue that if two lanes would be adequate for passenger cars the third and fourth are incremental to trucks, as that if two are adequate for trucks the third and fourth are incremental to passengar cars. In such a case all lanes are joint costs. Capacity costs can be regarded as incremental over time, however, and if the last lane is used only by peak traffic there is a case for charging its cost to peak traffic, where this is administratively possible.

The incremental cost principle is usually advocated as a method of making heavy trucks pay their share of costs. It is equally applicable in other cases, however. If heavy trucks do not use minor roads, these are incremental to light traffic and heavy vehicles should not contribute to their cost. Roadside beautification is not required by trucks and is largely for the benefit of passenger traffic. It can therefore be charged to passenger vehicles only, as an incremental cost. Good high-speed alignment and superelevation are similarly incremental to high-speed passenger cars.

Thus we conclude that the incremental cost principle is valid within its field, but this field is much more limited than is often supposed. Data are difficult to collect, and genuine cases of incremental costs difficult to identify. Where an item of costs can be assigned to a certain group of vehicles this principle does not help in the problem of distributing the cost between individuals in that group. Finally, it is no guide whatsoever to the allocation of joint costs, which remain the bulk of highway overheads.

Conclusions on Taxation Principles

The ability-to-pay principle is invalid for the allocation of costs between sectors and within the motor vehicle sector, but it might be used as a principle of general taxation for allocation of the community's share of costs among individuals. The benefit principle is completely unacceptable for allocation of costs either between sectors or within sectors. The amount-of-use principle has severe limitations and is practicable only for the allocation of overhead costs between vehicles within the vehicle sector. It should then be based on availability for use. The incremental cost principle is completely valid for cost allocation both between sectors and among vehicle types within the vehicle sector, though it cannot allocate costs among individual vehicles of one type. It can be used only to allocate costs which are genuinely incremental, however, and cannot allocate joint costs. Since most costs are joint it is therefore of limited usefulness.

A System of Economic Pricing for Highways

We have examined the popular principles of highway pricing and found them all either invalid or inadequate. Conclusions drawn from this examination are, however, of great value in devising a more satisfactory system. It has been found, for example, that it is desirable to make vehicles pay the marginal cost of each trip so as to discourage users from making trips which are not worth while, but undesirable to charge more as this will discourage trips which are worth while. It is also undesirable to make charges on the community which are not made by other industries in analogous positions, as this upsets the balance of economic competition.[11] It is quite equitable to allocate to certain groups of beneficiaries the costs of facilities from which they alone benefit, and there remains a field of benefit to motor vehicles, which may be shifted to property, which can be taxed as necessary and where convenient.

In using these conclusions to build up a pricing system we must be guided by the objectives which such a system of pricing should aim to achieve. The two basic functions of prices are to control supply and demand and to bring them into equilibrium. Uneconomic supply will be avoided in the case of a public utility by use of the planning theory outlined in chapters II to V, but it is still desirable that highways should pay their money costs with neither profit nor loss, because there is no inherent justification in using highways to achieve the transfers from one sector of the economy to another which would result from profit or loss. Sumptuary taxation may be applied to highway transport as to any other good for external reasons, but this is outside the scope of a pricing policy. Demand for highway services is in no way controlled by the planning analysis, however, and can be maintained at an optimum level only by use of direct controls or pricing policy. In the absence of direct controls one basic aim of the pricing policy should be to encourage the optimum volume of traffic on all roads. The pricing system must not differ in its method of charging for community services from the system used for other industries, especially other forms of transport, so as to maintain economic competition. Finally, where alternative arrangements would satisfy these aims, the choice should be made on grounds of equity. This will be largely a matter of opinion and is a suitable field for political decisions, provided this is limited to cases which equally satisfy the other objectives. The views expressed in these situations below are entirely those of the author. They are claimed to be reasonable, but different opinions held by other people might be equally reasonable.

Charging Direct Costs

To achieve optimum traffic volumes every vehicle should be charged the marginal cost of every trip. Marginal cost, however, includes vehicle costs and users' personal costs. Each vehicle unit will already bear average vehicle and users' personal costs and the pricing system need therefore take account only of the excess of marginal cost above average vehicle and users' personal costs. This is considerably more than is often imagined and is composed largely of the extra costs caused to other traffic by increased congestion. In practice it will be impossible to charge a different fee per vehicle unit mile on each road at each time, though this might be done in special cases such as bridges and tunnels where marginal costs are much higher at peak times. A certain amount of averaging will be unavoidable and will result in that part of the price structure which varies with use covering the average of the excess of marginal costs over average vehicle and users' personal costs on all roads in the group. If marginal cost is borne by each vehicle unit under conditions of increasing costs the authority will have revenue from this source greater than its total costs. This would tend to suggest that some expansion of the facility is called for, but rising costs alone do not necessarily mean that expansion is worth while unless the resultant reduction in costs would more than cover the capital outlay. In cases of decreasing cost the authority might[12] derive revenue from this source less than its total costs. Marshall and Pigou among others have advocated using the profits in increasing cost cases to subsidize decreasing cost cases. This is one of the most controversial issues in welfare economics and we are fortunately absolved from the responsibility of deciding its validity here by the practical impossibility of collecting exactly marginal cost on each road. The use of averages will mean that some roads will 'earn' more than their costs and others less, but this is defended not because it is theoretically desirable but because it is in practice unavoidable.

The two most serious objections to marginal cost pricing by public utilities are overcome in the case of highways.[13] The absence of a market criterion for investment is made good by the planning analysis of chapters II to V, and the necessity for some form of subsidy is avoided by some form of supplementary pricing system.

Over the highway system as a whole, charging marginal cost to traffic will cover more than average variable costs, but probably less than average total costs. The average variable cost curve will be constant or rise throughout its length and marginal cost must therefore be equal to or greater than average variable cost. If it is greater in any one case then total receipts from marginal cost pricing must exceed total variable costs. Where marginal costs exceed average total cost, however, there is an *a priori* case for expansion of the facility. This may be outweighed by costs in some cases, but if the system as a whole derives revenue from marginal cost pricing greater than total costs it is almost certain that there is scope for further investment. Marginal costs on English highways under current conditions will be high because congestion is considerable, and application of this principle might well result in a surplus. The conclusion that there is scope for further worth-while investment in English highways would hardly be a surprise, however. With an optimum highway system there will remain some part of total costs not covered by marginal cost pricing. This must be collected by taxes which do not vary with the amount of highway use, otherwise they will upset the marginal conditions.

Charging Remaining Overheads

As we have seen above it is equitable that genuine cases of incremental costs should be charged to the groups concerned. The central government will therefore be charged with the extra costs and sacrified benefits involved in any plan that was adopted because of the inclusion of community benefits such as defence needs and unemployment relief, over and above those costs involved in what would have been the optimum plan from other considerations alone. Similarly local governments will pay the extra costs of such services as sidewalks and sewers, it being assumed that the benefits from these are evenly distributed throughout the local community. A tax on pedal cycles to cover the cost of cycle tracks would be justified where these are provided. In charging sidewalks incrementally we must avoid allocating the common cost of right of way incrementally. The outer few feet of right of way which are used for the sidewalk might be the most expensive, but the sidewalk right-of-way cost would be less if the road were narrower. This is a case of common costs and cannot be allocated incrementally since the sidewalk is not added to the road any more than the road is added to the sidewalks. The historical order in which they were built is also irrelevant. The best solution here is to charge the local community for the proportion of right of way actually occupied by the sidewalk at its average cost.

Some costs can be allocated incrementally to groups of vehicles—for example, parkways to passenger cars—and in some cases strong surfaces to heavy vehicles. The community must contribute the full cost of its incremental services, but motor vehicle users have already made a contribution to their overheads by paying a variable tax equal to marginal cost, which is above average variable cost. Only a portion of fixed money costs will remain to be recouped if no profit is to be made, and it would therefore appear equitable to charge only this proportion of incremental costs to the relevant vehicle groups. This will be divided between individual vehicles within the group equally since it is an element of fixed cost which does not vary with the amount of use. It might be levied at a flat rate per vehicle, per axle, per wheel, per gross ton, and so on, but a convenient compromise among all these would be a flat rate per standard vehicle unit[14] applied only to the group in question. This could be collected as part of the registration fee.

The same proportion of remaining costs will have to be spread over all vehicles. One problem of equity arises here, however, in that average fixed cost will vary considerably between different types of road. It will generally be higher on residential streets with low traffic volume, or access roads as they are often called. We have so far rejected the idea that property derives a distinct benefit from access and should make a contribution to costs to represent this. But the fact that the property benefits are shifted vehicle benefits and that these benefits tend to shift to the sector taxed is a useful principle on which to separate the taxation of local and long-distance traffic. Since the level of average fixed cost still to be recouped will be higher on access roads than on main highways, and the access roads are incremental to local traffic, the users of access roads should pay a greater contribution to remaining costs than the users of main highways. The remaining costs on main highways can be expanded by the proportion of total vehicle unit miles to vehicle unit miles on main highways, and this amount allocated to all traffic. The additional amount which must be recouped on access roads can be charged to local traffic indirectly by means of a property tax. The level of property tax on sites served by each access road will be determined by the level of costs still to be recouped on that road. The benefit from which this tax is paid will be shifted from local traffic to property owners, and the incidence of the tax will fall on local traffic. The property tax is therefore a convenient method of charging to local traffic the higher average fixed costs on access roads; the indirect way of doing this being necessitated by the impossibility of identifying and charging local traffic directly.

Conclusions

The pricing system outlined above achieves many of the objectives for which it was designed. By charging for each trip at marginal cost it encourages optimum use of the highway plant. Where marginal cost reaches high levels in times of peak congestion a supplementary price in the form of a toll applied only at those times will discourage traffic which is not prepared to pay the full costs it causes, and thereby lessen congestion. The extent to which this is practicable will be limited, but bridges and tunnels are possible cases. Where peak traffic on crossings is reduced, congestion on approach roads will also be less. This toll might be arranged to include part of the peak marginal cost on the approaches as well as on the bridge to achieve this. By using supplementary charges to cover those costs not recouped from marginal cost pricing we are assured that the system as a whole will make neither profit nor loss. The supplementary pricing for remaining overheads is charged on availability for use and spread over all vehicle units so as not to upset the optimum position achieved by marginal cost pricing. Overhead costs which benefit only certain groups of traffic or the community are charged to them as far as possible. Property taxation is levied only on benefits shifted from that traffic which is responsible for the costs which the property tax is designed to meet.

The system therefore uses monetary incentives and disincentives as far as possible to achieve the optimum position. Some direct controls will be necessary on the maximum size and weight of vehicles, but these can be relaxed for exceptional indivisible loads, the charge for such relaxation being the marginal cost of the trip both to the highway authority and to the traffic which it inconveniences.

Perfection will not be achieved because of practical limitations. Averages of marginal costs on different roads and at different times will have to be used as a basis for the level of marginal cost pricing, and charges will not be at the ideal level in many cases. It will not therefore be possible to achieve optimum traffic volumes precisely. This is not a fault of the pricing system, however, but a limitation of the means of employing it. These practical difficulties are always present, but the system outlined above will still achieve as close an approximation to the optimum as is possible in practice. Its use will call for extensive data on costs, but these will already be available as a result of the planning analysis. With the use of modern techniques of handling data the cost of such a statistics and accounting service will be small by comparison with the value of the results achieved.

APPENDIX TO CHAPTER VIII

References on Pricing Principles

The letters on the left indicate the principles discussed, as follows:

A = Ability-to-pay Principle
B = Benefit Principle
C = Amount-of-use Principle
D = Incremental Cost Principle

AB H. Tucker and M. C. Leager, *Highway Economics* (Scranton Pa.), 1942.

ABCD Bertram H. Lindman, *A Proposed System of Highway Financing for the State of California* (Sacramento, 1946).

BC Wilfred Owen, *A Study in Highway Economics* (Cambridge, Mass., 1934).

BCD Public Administration Service, *Financing a Proposed Highway Program in Minnesota* (Chicago, 1954).

BCD Wisconsin Legislative Council, *Interim Report on Highway Finance* (Madison, 1952).

BCD M. Earl Campbell, "Considerations in the Assignment of Financial Responsibility" (1950).

BCD G. P. St. Clair, "Suggested Approaches to the Problems of Highway Taxation," H.R.B. Proceedings, 1947.

BCD C. A. Steele, "Information Needed for the Fiscal and Allied Phases of Long-Range Highway Program Planning," H.R.B. Bulletin no. 12, (Washington, D.C., 1948).

C Oregon State Highway Department, *Appraisal of the Highway, Road and Street Problems* (Salem, 1949).

C James C. Nelson, *Financing Washington's Highways, Roads and Streets* (Olympia, 1948).

CD Western Highway Institute, *Highway Taxation Problems: A Synopsis*, Technical Bulletin Series no. 3, 1950.

CD *Financing Needs and Allocating Costs of Highways Among Highway Users in Utah*, Bureau of Economics and Business Research, Utah University, for the Legislative Council of Utah (Salt Lake City, 1950).

D Bureau of Railway Economics, *An Economic Survey of Motor Vehicle Transportation in the U.S.A.* (Washington, D.C., 1933).

D Federal Co-ordinator of Transportation, *Public Aids to Transportation*, vol. IV (Washington, D.C., 1940).

D James W. Martin, "Initial Problems Confronted in the Kentucky Incremental Cost Study," H.R.B. Bulletin no. 121.

D Washington State Council for Highway Research, *Nature of Highway Benefits* (Olympia, 1954).

CHAPTER NINE

THE COLLECTION OF REVENUES

IN THE LAST CHAPTER we examined the principles on which the burden of the financial cost of highways should be divided among the beneficiaries. Now we must examine the devices by which each group's or individual's share can be collected. Just as it was useful to examine critically the advantages and disadvantages of the conventional principles of taxation as a basis of our own system, so is it useful to examine the desirable and undesirable features of the conventional forms of taxation[1] as a guide to devising a system which will collect each part of the total revenue by the best available means. There are three main groups of highway taxes: taxes on vehicle use which vary with the amount of use made of the highways; taxes on vehicle use which do not vary with the amount of use made of the highways; and taxes on bases other than vehicle use.

VARIABLE VEHICLE TAXES

Fuel Tax

The gasoline or fuel tax has become one of the most widely used and most popular means of collecting revenue for highways, and often for other purposes as well. It has all the fundamental advantages of a good tax, by comparison with which its faults are normally regarded as unimportant. It is popular, both with the collecting authority and the taxpayer, because of its simplicity. A fixed amount per gallon of a sufficiently standardized commodity is easy to reckon, collect, and control, and administrative costs are therefore a very low proportion of total revenue. It is paid in small quantities at frequent intervals by the consumer, and regarded by him as part of the price of fuel, so that payment of the tax becomes relatively painless.

From the theoretical standpoint it measures use of the roads in a rough way. For a given vehicle it is directly proportional to mileage, and it increases per mile with vehicle weight, though less than proportionally. This relationship to weight is one of the limitations of the fuel tax as it means that the heavy vehicle, which is normally considered to cause greater highway costs per gross ton-mile than the lighter vehicle, pays less in fuel tax per gross ton-mile. Fuel consumption varies with speed in a U-shaped curve. At low speeds consumption is high, and low speeds are usually brought about by congestion, when marginal costs are high. In the medium-speed range both fuel consumption and marginal costs are at their lowest, while at high speed fuel consumption increases, and so does the impact load and consequent wear and tear on the highway. Thus the fuel tax, in being directly proportional to consumption, is a function of mileage, weight, and speed. It varies proportionally with costs as a function of mileage, and in the right direction if not to the right extent with weight and speed.

Fuel consumpion also varies with the type of road on which the vehicle is operated, being higher on poor surfaces. Poor surfaces, however, cost less to construct than good surfaces. This is often claimed as a case where the fuel tax varies inversely rather than directly with highway costs. If the tax is used to recoup marginal costs alone, however, this does not apply, since the costs of highway wear and tear for a given vehicle-mile are higher on a poor surface than on a good surface. Thus although the vehicle derives less benefit from the poor surface it occasions higher direct costs on the highway authority, and that part of pricing which varies with use should therefore be higher. Whether a low-type or high-type surface is best on a given road is largely a function of the volume of traffic, and the advantage of a good surface to the user is therefore, in a sense, an economy of scale. Where the economy is not possible because demand is low, and marginal highway costs are therefore high, the marginal price should likewise be high. Thus the fuel tax varies the right way as a function of surface type, though again not necessarily to the right extent. This defence of the fuel tax on this point is quite distinct from that often advanced that the worst roads should collect most tax because they are most in need of improvement. It does not follow that the roads with the low-type surfaces should always be improved, nor that users before the improvement should pay for it. This argument is therefore invalid. We are assuming here that as a result of use of the planning analysis of chapters II to V all roads have the optimum type surface for their anticipated traffic volumes.

The indirect incentive and disincentive effects of the fuel tax are often regarded as faults. Where it is used to recoup only those highway costs which vary with use, however, these indirect effects are advantageous. It will result in less use of existing vehicles than if fuel is not taxed but the reduction will involve only those journeys which are worth an amount between full operating cost without fuel tax and full operating cost with fuel tax. Where fuel tax reflects only marginal cost of highway use, these trips are worth less than their full costs and it is therefore economically advantageous if they are not undertaken. A fuel tax might also reduce demand for vehicles, or shift demand to those vehicles with lower fuel consumption. Again, where it reflects only marginal highway and congestion costs of vehicle use, it will reduce demand for vehicles only by that amount which is not worth its full costs. If a vehicle with high fuel consumption inflicts higher marginal highway and congestion costs than one with lower fuel consumption, and this is just reflected in the fuel tax, it is economically advantageous that any person who does not consider the advantages of the vehicle with the higher fuel consumption worth these extra costs should shift his demand to one with lower fuel consumption. Thus, while the fuel tax reflects marginal costs of highway use, no more and no less, it will have the desired incentive and disincentive effects. A higher or lower fuel tax will have undesirable incentive or disincentive effects, however, which is the very reason why it is not appropriate to use a tax which varies with highway use to collect those costs which do not.

One difficulty with fuel taxation arises with vehicles using fuels other than gasoline. At present gasoline and diesel fuel account for effectively all motor vehicle traffic, but the rates of fuel consumption of gasoline and diesel powered vehicles differ considerably. A survey in the State of Washington showed that intra-city diesel buses covered 58 to 62 per cent more miles per gallon than similar gasoline buses; inter-city buses achieved 52 to 78 per cent more and trucks 15 to 71 per cent.[2] A higher tax per gallon is therefore needed on diesel fuel than on gasoline if users of similar vehicles are to pay the same tax per mile. This can be achieved by basing the ratio of the two taxes on the weighted average ratio of consumption. It is sometimes argued that the diesel tax should be higher than this ratio would suggest so as to tax the users of heavy vehicles more than proportionally with their fuel consumption. This would be a sound argument if all heavy vehicles were diesel-powered, but while the two power forms are alternatives over such a wide range of vehicles, discrimination against diesel would encourage the uneconomic use of gasoline engines, just as discrimination against the gasoline engine by equal tax rates per gallon encourages the uneconomic use of diesels. The only way to avoid these undesirable effects is to make the tax the same for either form by levying a rate per gallon on alternative fuels inversely proportional to consumption. The greater administrative problem with the taxation of diesel fuel, caused by the higher proportion of total consumption used for non-highway purposes, means that collection costs will be a higher proportion of total revenue than in the case of the gasoline tax. Collection costs will not be affected by the rate of the tax, however.

Thus we conclude that the fuel tax is a very useful method of collecting that part of the total price for highway use which should vary with the amount of use. Its great limitation is that it discriminates in favour of heavier vehicles, which cause greater variable costs per gallon of fuel consumed than lighter vehicles. Furthermore, it takes no account of the number of axles or wheel loads, which can be important factors in determining marginal highway costs. We therefore need some alternative or supplementary tax which will take account of these items.

Tire Tax

In some countries tires are taxed, but usually at a flat rate by weight. This has some of the advantages of the fuel tax but the additional disadvantages of encouraging the prolonged use of worn tyres with a consequent reduction in safety. This tax could be applied, however, so as to have a great advantage over the fuel tax. We cannot distinguish the fuel used in heavy vehicles from that used in light vehicles, but heavy vehicles use larger tires and these can therefore be distinguished. A tire tax, steeply progressive with size, would discriminate against heavy vehicles, and could therefore be arranged to offset the failing of the fuel tax which discriminates in their favour. With the two taxes the net effect could be graduated with total weight as desired, in relation to the variable costs incurred by different weights.

The indirect effects of such a tax would mainly not be detrimental. It would encourage the use of smaller tires and might therefore have an effect on vehicle design. However, if the tire size on a given vehicle is reduced, more wheels will be needed to spread the load. Assuming total tire wear in terms of weight of rubber to be constant irrespective of the number and size of tires,[3] the vehicle with more and smaller tires will pay less tire tax. The effect of a progressive tire tax on vehicle design is likely to be slight in practice, however, because savings on tire tax would normally be more than offset by the additional production cost of vehicles and tires. It is not known exactly how highway wear varies with the number and size of tires for a given vehicle, but as data on this become available the tax could be suitably adjusted. Thus, it might well be found that highway wear is inversely related to the area of tire in contact with the road, and that the use of wider treads would reduce highway wear. This could be allowed for by charging a lower rate of tax on wide tires. To do this, while retaining the higher rate on larger tires so as to levy a higher tax on the heavy vehicle, would call for a more complicated system than a simple scale progressive with weight. Taxation might then be made steeply progressive with diameter, but regressive with width. Thus, the bigger the diameter, the higher the tax per pound of rubber, because heavy vehicles use bigger wheels; but the wider the tread the lower the tax per pound of rubber, because wide tires inflict less damage on the highway than narrow tires. These different factors could be combined into a schedule of the tax rate for each size of tire. Whatever the relationship between tire size and highway wear, the schedule could be adjusted so that persons using tire arrangements which inflict greater damage would pay correspondingly more in tire tax. It is the task of further technical research to establish what these relationships in fact are.

The usual objection to tire taxation is that the rate of tire wear varies with many factors, especially the road surface. Tire wear is a function of impact force, however, which in turn is a function of many factors. It is a law of dynamics that the road hits the tire just as hard as the tire hits the road, and tire wear will therefore vary directly with wear and tear of the road surface. As with fuel consumption, operating costs will be higher on poorer surfaces, but so will marginal costs of highway wear, justifying higher tax rates. Stop-and-go conditions increase tire wear, but these are usually caused by congestion when marginal costs of highway use are high.

Thus we conclude that a judicious use of a fuel tax combined with a suitably based tire tax, can be made to vary approximately with the marginal costs of highway use, allowing for all types of vehicles. Two other forms of tax which vary with use have received widespread support in the United States, however, and deserve examination.

Weight-Distance Taxes[4]

These are imposed on heavy vehicles by some states in the U.S.A., and vary round a basic form. Some are based on gross weight, others on declared maximum operating gross weight, net weight, load, etc., or on receipts for common or for-hire carriers. Vehicles are usually grouped and a set fee per mile for each group levied on the basis of the incremental costs attributable to that group, and the average weight of vehicles in the group. Oregon is the leading exponent of the tax in the United States and uses a system of 2,000 lb. groups for common, contract, and private carriers over 6,000 lbs., with the rate for each group based on average anticipated maximum gross weight. Allowance is made for fuel tax paid and public and farm vehicles are exempt. A daily record of each vehicle is kept by the operator on a prescribed form, and these are summarized monthly in the report on which the tax is based. Unsatisfactory reports are subject to assessment with the right of appeal to a public hearing.

The advantages of such a system are considerable. It allows taxation to be levied on those vehicles which are actually responsible for costs and can be limited to, and made progressive on, heavy vehicles, with gross weight or wheel load as the tax base. By charging each vehicle according to the highway costs it causes, this tax discriminates neither against nor in favour of the heavy vehicle, and thereby allows full advantage to be taken of its natural advantages without subsidizing it at the expense of lighter vehicles. It thereby overcomes the difficulties of the fuel tax which takes inadequate account of weight, and the registration tax which takes no account of mileage. Such a flexible tax which can be used to place the burden of cost wherever the authority feels it should fall is naturally attractive. It also has the great advantage in the U.S.A. of taxing traffic originating in other states. Out-of-state vehicles may pay no registration fee and purchase no fuel in a state through which they pass, but they are required to make mileage tax returns in the same way as intra-state vehicles, and are subject to checking at weighing stations, which are usually located near state borders. This is particularly important for those states lying on trunk routes between termini, the so-called "corridor" states.

The weight-distance tax is not without its difficulties, however, both theoretical and practical. Most of the so-called theoretical objections are to particular applications of the principle and could be overcome. As a straight ton-mile tax it takes no account of wheel load or space requirements, but it could be based on either of these. Weight groups are necessarily arbitrary but some averaging is necessary with any highway tax. In some states it applies only to common carriers, but this discrimination is not fundamental. If used only to recoup those costs for which heavy vehicles are responsible it does not penalize the greater efficiency of heavy vehicles, but puts them on an equal economic footing with lighter vehicles. The greatest single theoretical objection to the tax as it is usually used is that it recoups non variable costs on a variable basis. Costs which are incremental to heavy vehicles are largely "overheads" in the sense that they do not vary with the amount of use by such vehicles once they are incurred. To this extent they are not marginal to each trip, and if charged on a trip or mileage basis some traffic will be discouraged, although it would be prepared to pay its full marginal costs. In the United States this tax is used to recoup these non-variable costs. If it were limited to that part of incremental costs which varies with each trip—that is, truly marginal costs—the amount of revenue collected would be very small.

The tax has many practical difficulties. In some states it has been declared unconstitutional on grounds of discrimination, while in others it has seriously threatened reciprocity agreements with neighbouring states. It is extremely difficult to administer and enforce and is unpopular with those taxed because of the inconvenience and clerical work involved. The use of sealed meters is impossible because only mileage within the state is recorded and the system therefore has to rely on taxpayers' reports. These are difficult to check for non-scheduled services, enforcement costs are high, and it has therefore come to be called a tax on honesty. Revenue collected is small, being nowhere more than 1 per cent of total highway revenues,[5] and would be negligible if limited to marginal costs. Collection costs average 18.7 per cent of revenues in the U.S.A. by comparison with 0.75 per cent for fuel tax and 9 per cent for registration taxes,[6] the latter being largely the administrative costs of motor vehicle departments which the registration fee is designed to meet. In North Dakota collection costs were as high as 47.7 per cent in 1947, and the low Oregon figure of 4.2 per cent results from excluding enforcement costs of police and weighing stations, and collecting considerably higher revenues than is justified on a variable cost basis. Twelve states[7] have instituted and discarded weight-distance taxes as legally or administratively impracticable.

The practical difficulties with this form of taxation are such that, though theoretically attractive, it is barely worth while. Collection costs are high in relation to revenue. They would be much higher proportionally if only marginal costs were recouped, and it is theoretically undesirable to recoup more than marginal cost by a tax which varies directly with use. It would therefore appear that any advantage which this tax might have in addition to a combination of fuel and tire taxes is so small that its use is not warranted. A weight-mileage tax on all vehicles could be administered by use of sealed odometers, where all mileage is to be taxed, in much the way electricity meters are used. Such a system could charge marginal costs as an alternative to fuel and tire taxation, but while it would be more precise in some respects than fuel and tire taxes, it would be much more expensive, and the rate of tax per mile would not reflect highway and traffic conditions and costs in the way that the fuel tax does.

Tolls

Tolls are normally used as a means of recouping full costs on a loan-financed project. As such they have the great disadvantage of discouraging marginal traffic by charging average total costs. It was estimated that 63 per cent of the traffic which would use the Ohio Turnpike if toll-free would be discouraged by the recommended toll schedule.[8] This traffic would in any case have paid normal fuel taxes and so on. If used to charge only marginal cost, tolls would be theoretically perfect in allowing each unit of traffic to be charged precisely marginal costs on each stretch of road at each time. It would, however, be administratively impossible, or at least highly inconvenient and very expensive, to have toll gates on every stretch of road charging comparatively low tolls. The idea of charging a specific fee for use of a certain section of road, or bridge or tunnel, at certain times may be very useful as a supplementary charging device, however.

Non-variable Vehicle Taxes

Registration Tax

This is levied per period of time for the right to operate a certain vehicle on public highways. It does not vary with the mileage of the particular vehicle, but can vary with almost anything else. Horsepower, size, weight and type of vehicle are often used as bases of taxation. It is a very popular method of recouping the administrative expenses of the licensing authority and is often high enough to make some contribution to the fixed costs of highways. Its great advantage for the latter purpose is that by not varying with the amount of use made of the vehicle it has no disincentive effect on the marginal trip. If overhead costs were split evenly between vehicles on this basis, however, the rate might be so high as to discourage the marginal user altogether.

It is often held that although joint costs cannot theoretically be allocated between vehicles it is still equitable that persons using the highways most should make the largest contribution. The use of vehicle classification makes this possible. A tax based progressively on weight puts a greater burden on heavy vehicles. These have higher operating costs and a flat fee would therefore be a smaller proportion of total costs for a given mileage, and less likely to discourage use altogether than in the case of a light vehicle. Use of a graduated scale relieves the burden on the light vehicle where it is more likely to discourage use, and increases it on the heavy vehicle which is more likely to be prepared to meet it. Part of the higher fees on heavy vehicles can be justified on an incremental basis. As we saw in the last chapter some highway costs can be isolated as being incurred for the benefit of certain groups of vehicles only. These costs can justly be allocated to those vehicles, but graduation of registration fees beyond this is a clear case of price discrimination.

Amount of use is more difficult to incorporate into a registration fee as we are faced with a paradox. On the one hand it may be felt that the person using the highway most should pay most, but on the other hand it is undesirable that the contribution of any person toward overhead costs should vary with use, as this discourages the marginal trip. A compromise can be achieved to a certain extent by use of averages for vehicle groups. Thus, if the average annual mileage of common carriers is higher than that of private carriers, the registration fee for all common carriers may be higher. The amount paid by any individual will not vary with his own mileage, but in the long run with that of the group to which he belongs. This argument is often advanced to justify lower rates on farm vehicles which cover low annual mileages, though the motive one suspects is more often political.

An interesting form of discriminatory tax is levied in Paraguay on private cars. This varies with the make, size, and age of the car, being steeply progressive with value. It cannot be justified by a claim that the more expensive cars use the roads more, but accords with the ability-to-pay principle of taxation on the assumption that more valuable cars are owned by wealthier people. Its great attraction from the economic standpoint is as a means of price discrimination. If it is aimed to collect a given revenue from registration fees and this is spread equally over all cars, the poorer motorist might give up motoring altogether. This is avoided if a low rate is charged on cheap cars and compensated by a higher rate on more expensive models. A given amount is a lower proportion of total operating costs on an expensive model and is therefore likely to have less disincentive effect, especially since the owner of the more expensive model is probably wealthier.

The indirect effects of discrimination in registration fees whether by weight, class or value will probably be undesirable since there is no economic justification for them. Generally the effect will be to switch demand to the group with the lower rate. Some contract carriers may operate as private carriers by fictitious purchase and sale of cargo, some operators may use lighter vehicles, and some motorists cheaper cars. The flat registration fee for private cars in the United Kingdom was aimed to encourage demand for, and production of, higher horsepower models which would compete more effectively in export markets. This effect has been largely offset by increased fuel taxation, however, and the effect of a graduated registration fee in diverting demand to smaller cars was probably overestimated anyway, as the difference in fee was but a small proportion of total costs. The extent to which the lower-taxed vehicle will serve the purpose of the vehicle previously used is so limited that shifts in demand will probably be of minor importance. There remains the equity problem which inevitably arises with grouped data. One individual common carrier might cover a lower mileage than one individual private carrier and one large-car motorist might have a lower income than one small-car motorist. In practice a compromise must be drawn between the undesirable effect of the flat fee of discouraging the marginal user, and the alternative undesirable consequences of any discriminating system.

Other Non-variable Taxes

Certain minor taxes sometimes levied will have much the same effect as the corresponding parts of the registration fee. Fees on carriers' licences are effectively part of the registration fee. Drivers' licence fees will have much the same effect as the flat registration fee. There seems little point in using these forms of taxation to contribute to highway costs, but it might be convenient, as is often done, to levy them at low rates to cover the administrative costs of the controlling authority.

Personal property taxes on private motor cars may be levied for one of two purposes; as a source of highway revenue and as a source of general revenue. The former will have precisely the same effect as the registration fee graded with value, and there is little point in using both. Where personal property taxation is used for general purposes it should apply to motor cars no more and no less than to other durable goods. Its disincentive effect will then be negligible, since if a person gives up his car he will still pay personal property tax on whatever he buys instead.

Purchase tax levied on new vehicles only will vary with value and be treated as part of the initial cost to be depreciated over the life of the vehicle. It will be passed on to later buyers to an extent roughly proportional to the remaining value. As such it will have much the same effect as a registration fee graded with value, a purchase tax of x per cent on initial price corresponding to a registration tax of x per cent of the annual depreciation of the car, itself a function of value. To be precise, interest should be allowed for on the purchase tax as this is levied on the car's whole life at the initial sale, whereas an annual tax spreads the burden over time. Further, the purchase tax per annum in being a function of depreciation in value is in part variable with mileage. That one is sometimes levied initially on the dealer and the other on the owner is unimportant as the incidence of both will be shifted partly to the other, depending on the rate of tax and relative elasticities of supply and demand. Since the effects and incidence of both are roughly the same there is little point in using more than one of them. Again, purchase tax on cars might be used at the same rate as on similar goods for general purposes.

The sales tax often differs from the purchase tax in being levied on second-hand car sales too. As a source of highway revenues this is a disadvantage as it discourages otherwise desirable transactions for no good purpose. Given the amount of revenue to be raised, the purchase tax reduces net private consumption by that amount; the sales tax reduces consumption by the same amount but also reduces transactions on second-hand goods. Since these transactions would have yielded a net benefit, otherwise they would not have been undertaken, this represents a net loss. It would therefore appear that the sales tax which applies to second-hand dealings too is inferior to the purchase tax which does not.

Non-Vehicle Taxes

General Taxation

Various forms of taxes are used by central and local governments to raise revenue for general purposes—income tax, death duties, sales and purchase taxes, poll taxes, and so on. The relative merits of these as general sources of revenue are beyond the scope of this book. Those parts of highway costs which can justly be allocated to the central government for defence and unemployment relief or local governments for sidewalks and slum clearance, for example, are similar to other forms of government expenditure and will be met from these general forms of taxation. As such they are not specifically tied to highways and do not come within an economic pricing system.

Property Taxation

Having discarded the idea that property derives any benefit from highway improvements distinct from shifted benefits to traffic, we are left with three possible uses for property taxation. The first is as a general source of local government revenue. This is one of the commonest uses of the property tax and there is much to be said for assessing one's liability for local government costs on the basis of the property one occupies; it is evenly spread, simple to administer, and graduated very roughly with wealth and one's interest in local services. Where this source of local government finance is used it is an appropriate source from which to recoup the local government contribution to highway costs.

There remain two forms of property tax tied to highway use which are intended to recoup part of the motor vehicle share; general assessments and special assessments.

A general assessment on all property for highway purposes would shift the burden from the motor vehicle user to the community. Not all property benefits directly from highway improvements, and even taking highways as a whole the benefits which are shifted from motor vehicles to property are very unevenly spread. It would be an unjustified assumption that motor vehicle benefits are spread in proportion to property values over the country as a whole, and a general property assessment is therefore inappropriate.

Special assessments can be on property lining individual streets or on defined districts. We are not concerned here with special once-for-all assessments of property, say for paving purposes, but those taxes aimed to recoup from property owners the costs of the road attributable to the traffic which has conferred benefits on that property. As we saw in the last chapter, average total fixed costs can be higher on access roads than on main roads by reason of low traffic volumes. The costs of these roads are largely incremental to traffic using them and the property tax is a useful way to isolate benefits to this traffic and tax them after they have been shifted to property owners.

The special assessment of a district is an extension of this principle. Major highway improvements bring increased property values in their path. Normally the overhead costs of such projects are paid from vehicle taxes but there are here two difficulties: if the tax varies with use it discourages the marginal trip, and if it does not it discourages the marginal user. Vehicle taxes could be at lower levels if part of the costs were recouped by taxation of the property to which a large part of the vehicle benefits is shifted. There would be no disincentive effects attached to the property tax for, as we saw in chapter IX, benefits tend to shift to the extent of the propery tax. The benefit shifts most from those vehicles with most benefit to shift, so that in no case is a vehicle discouraged from operating while it can yield any net benefit. In effect, therefore, a special property tax on areas to which new highways have brought enhanced property values is a method of discriminating between vehicles, putting the biggest share of the burden of overhead costs on the vehicles with the greatest amount of net benefit to bear it, and avoiding discouraging the marginal user by excessive taxation.

The extent of this field for taxation can be seen from the data in table X for four parkways in New York.

TABLE X

EFFECT OF PARKWAY CONSTRUCTION ON LAND VALUES

	Value at time of construction $	Value in 1953 $	Percentage increase
Grand Central Parkway Constructed 1925			
Influence area	4,359,970	93,210,860	2038
Control area	141,540,460	850,806,405	501
Henry Hudson Parkway Manhattan Section Constructed 1935			
Influence area	397,500,370	415,868,110	4.62
Control area	1,484,218,230	1,367,626,890	− 7.86
Henry Hudson Parkway Bronx Section Constructed 1935			
Influence area	21,926,350	44,300,390	102.04
Control area	99,230,575	76,531,960	−22.81
Shore Parkway, Brooklyn Constructed 1939			
Influence area	61,695,802	108,710,361	76.2
Control area	487,750,170	581,496,483	19.2

SOURCE: "Effect of Public Improvements on Property Values," Office of the New York City Construction Co-ordinator, 1954.

The maximum annual sum that could be collected would be the current rate of interest on the difference in capital value between the present value of land in the area, and what the value would have been without the highway, as shown by the control area. In practice no collection system could achieve more than a small part of this, but the sums involved are such that this is still a valuable source of funds to meet the heavy overhead costs of major highway projects.

An alternative system of taxing this same rise in property values is the procedure known as recoupment, by which the authority buys land at the low price before the improvement, and sells it at a higher price afterwards, thereby deriving a once-for-all capital gain to pay part of the highway construction cost. This has a great advantage in encouraging development as the land starts again free from restricting leases, and can be redivided into more convenient sites, thereby avoiding the problem of remnants from sites only part of which have been used by the highway. This applies only along the route of the highway itself and will be less important with a limited access project. The disadvantages of the system are the risk involved in not knowing what the change in value will be, the considerable administrative and legal costs of buying and selling, and the cost of waiting while the area develops. It may take many years for the full increase to occur, and this will be hindered if the highway authority refuses to sell land. If it does sell, it will do so at less than the value to which the land will eventually rise. By leaving the property in private hands and assessing it for special taxation, however, receipts can be geared to development.

A Proposed Pricing System

Having examined in the last chapter how the burden of highway costs should be spread, and in this chapter the tools available for spreading it, we must now outline the procedure by which the highway authority might decide what part each tool should play. In the planning stage of the analysis each section of highway was considered separately. It is not possible to isolate each section in the pricing analysis, however, and we must therefore work largely on averages, correcting exceptional cases of deviation from the average by special taxes where possible.

In chapter VII the method of determining the annual money cost of the highway system to the highway authority was outlined. This is the total which must be collected in revenue, and the procedure for building up a pricing system is essentially one of deducting the contributions from each tax in order of desirability until the full amount is collected.

First comes the principle that marginal costs should be charged for each trip. The analysis of chapter IV determined the optimum volume of traffic and optimum level of marginal cost per vehicle unit on each road. From this must be subtracted the vehicle costs and users' personal costs to leave the net amount which must be collected by the tax. The weighted average per vehicle unit per mile of these net marginal costs on all roads at all times will be the amount to be collected by taxes which vary directly with highway use. The main tool for collecting this is the fuel tax, but because this charges the heavy vehicle a lower rate per vehicle-unit-mile[9] it will need to be supplemented by use of a suitably scaled tire tax, the two together levying a standard rate per vehicle unit mile.

On some roads at some times this average figure will be well below the prevailing marginal cost and the difference might here be collected by a toll, the object being either to discourage altogether traffic which is not worth its marginal costs, or to divert traffic to an alternative route where marginal cost is lower.

Most large cities have daily peak traffic periods, when marginal costs are considerably higher than at other times by reason of the delay caused to other traffic by each additional unit. Fuel consumption and therefore fuel tax will be higher in congested traffic, but the increase will be much less than the increase in marginal cost. The effect is that all traffic in peak times pays less than marginal cost and the value of some of this traffic is below full marginal cost. Removal of this submarginal traffic would considerably reduce congestion and increase net benefit. This can be achieved by levying a toll at peak times only, to bring variable taxation up to the level of marginal cost. Bridges or tunnels across rivers or railways would be suitable points for toll stations, the exact number and location being suited to the individual case, with the aim of intercepting most traffic once. Most of the traffic at peak times will be commuter traffic and therefore follow regular routes. The toll on a bridge can be enough to fill the gap between existing taxes and marginal costs not only on the bridge but on the average distance of travel in the congested area for all traffic crossing the bridge.

The appropriate level for such a toll is shown in diagram 16, where average operating cost and marginal cost are the cost curves for the average total dis-

Diagram 16

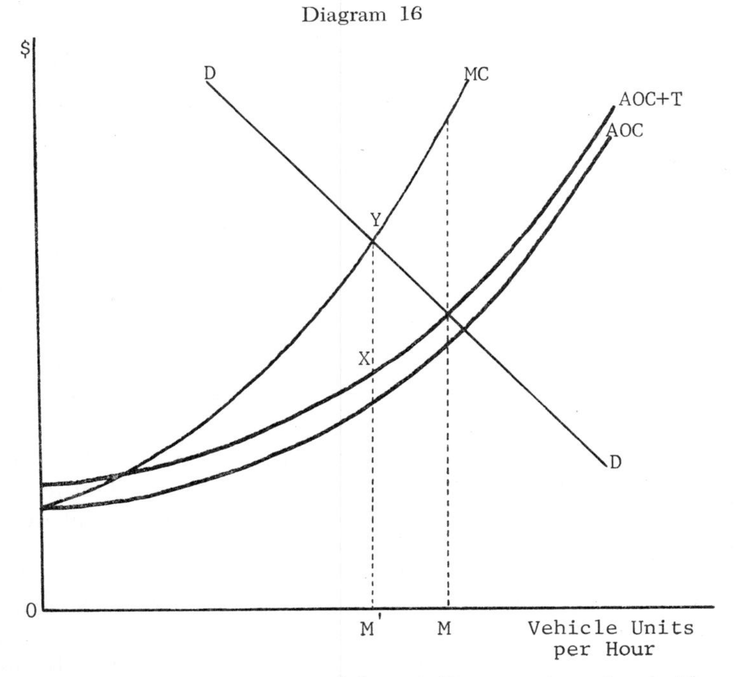

tance in the congested area covered by traffic crossing the bridge. Average operating cost includes only the variable vehicle costs and users' personal costs. To this must be added fuel and tire taxation (*T* per vehicle unit trip) to give the average variable cost to the traveller. The line *DD* is the demand curve, and *OM* journeys will be undertaken since this is the number worth more than the cost to the user. *M'M* trips are therefore undertaken although marginal cost is above demand price. To limit traffic to the optimum level *OM'* the cost to the traveller at *OM'* must be increased from its present level *M'X* to the demand price for this volume *M'Y*. A toll of *XY* per vehicle unit is therefore needed.

Where there are two alternative routes between two points traffic will spread itself so that average cost to the user is equal on both routes. The optimum position would be where marginal costs are equal, and this could be achieved by use of differential tolls. This is illustrated by a simple case in diagram 17. A fixed volume of traffic *OO'* divides itself between routes *X* and *Y*. The average operating and marginal cost curves for each are shown, those for *X* plotted from

Diagram 17

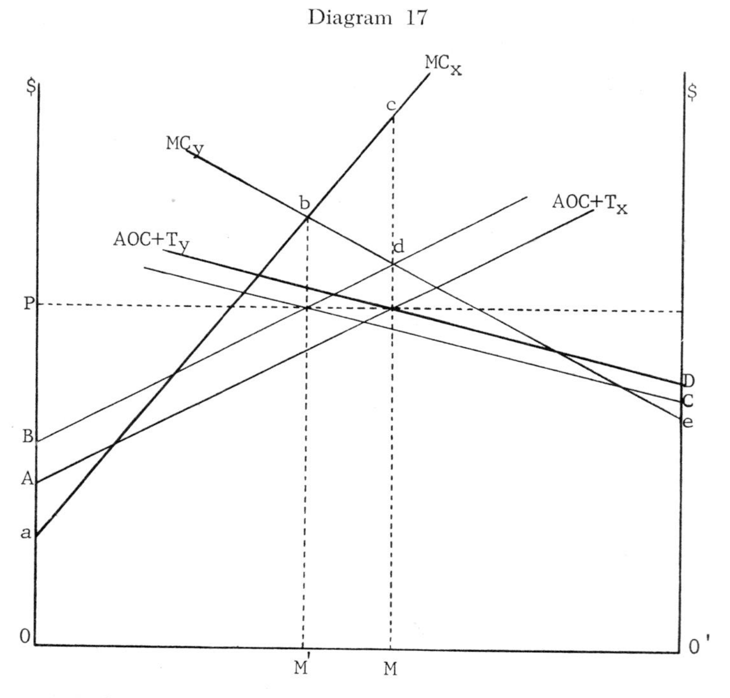

the left axis and those for *Y* from the right axis. The line *AOC* + *T* includes vehicle and users' personal costs plus fuel and tire taxes. Traffic will divide itself so that average costs are equal, i.e., *OM* on *X* and *O'M* on *Y* at *AOC* + *T* = *OP*. The optimum position, however, would be where marginal costs are equal, i.e., *OM'* on *X* and *O'M'* on *Y*. This could be achieved by a toll of *AB* on *X* and a negative toll or subsidy of *CD* on *Y* so as to leave *AOC* + *T* equal to *OP*. Total costs are the combined areas under the marginal cost curves, *acde*, where traffic is divided at *M*, and *abe* where it is divided at *M'*. Thus by imposition of toll and subsidy, on the assumption of no collection costs, the cost to the traveller is *OP* for the trip as before, but since total costs are less by *bcd* the highway authority will have a surplus of *bcd* between receipts from tolls and the cost of the subsidy. This hypothetical case proves that by redistributing traffic so as to equalize marginal rather than average costs a surplus can be achieved while no one is worse off.

It does not follow that the aim should be to keep *AOC* + *T* and the volume of traffic constant. Diagram 18 illustrates the position where there are two routes through a town, route *X* through the town centre and route *Y* a bypass.

Diagram 18

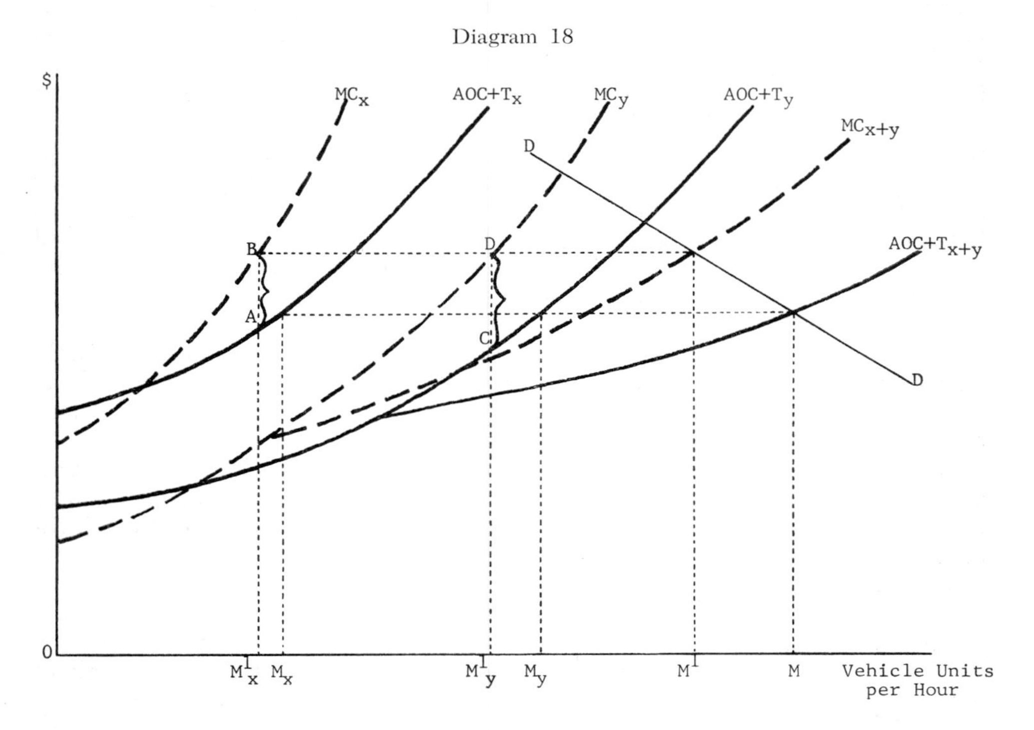

Each crosses a river by a different bridge providing suitable toll points. *AOC* + *T* again includes vehicle and users' personal costs plus fuel and tire taxes. The total curves are the horizontal addition of the individual curves. *DD* is the demand curve in the peak. With no tolls *OM* traffic will flow, divided into *OMx* on *X* and *OMy* on *Y*. The optimum will be *OM′*, the volume at which the demand curve cuts the total *MC* curve, divided so that marginal costs are equal on both routes, i.e., *OM′x* on *X* and *OM′y* on *Y*. This volume and distribution can be achieved by tolls of *AB* on *X* and *CD* on *Y*.

By use of this type of analysis, the correct tolls at each time of day can be computed for all the points where it is practicable and worth while having toll gates.

Receipts from fuel and tyre taxation and tolls, which together reflect the principle of marginal cost pricing, will be less than total money costs of the authority. The next contribution to costs comes from charging incremental costs to specific groups; the nation as a whole, the local community, classes of vehicles, and users of certain roads.

The central government will be charged for those costs incurred for purposes which are the responsibility of the community as a whole, but such contribution should be limited to those costs which are genuinely incremental to these purposes. Any highway project will yield some benefit for defence purposes, and will create employment, but this is true of most economic activities. The only circumstance under which part of the cost of a project is incremental to these purposes is where a plan is adopted by reason of these negative community costs being included in the analysis, and where a different plan would have been optimum had they been excluded. The incremental cost to society as a whole is composed of additional costs to the highway authority and vehicle users, minus any additional benefit to vehicle users, plus any reduction in benefit to vehicle users, which result from adoption of a plan other than what would be the optimum plan if benefits to society were excluded from the analysis. Where the inclusion of these benefits in the form of negative community costs influences the result of the analysis, the incremental cost to society must be positive, but less than the value of the additional benefits to society. The sum of such incremental costs on all highways would be paid by the national treasury to the highway authority annually.

Each local government would similarly make an annual payment to the highway authority to represent those incremental costs incurred as a result of the inclusion in the analysis of benefits to the local community such as the provision of sidewalks, assistance in slum clearance schemes, and so on.

When the contributions to highway costs by central and local governments are assessed in this way the various government departments will be encouraged to put a true valuation on community benefits which will be within their jurisdiction. Underevaluation might result in their interests being inadequately considered in the selection of the optimum plan, while the knowledge that they will be charged for any additional costs borne as a result of including the benefits in the analysis, up to a maximum of the value they place on such benefits, would discourage overestimating the value of benefits. The task of valuing community benefits would not then rest with the highway authority but with the appropriate government departments and ministers. Motor vehicle users would be disinterested in such decisions since the net benefit derived by them would be the same no matter what effect the valuation of community benefits has on the result of the planning analysis.

All remaining overhead costs are the responsibility of motor vehicle users, but some of these can be allocated incrementally to certain groups of vehicles, for example, some of the costs of thick surfaces to heavy trucks. Since the variable taxes have already recouped part of overhead costs, however, only part of the incremental costs should be allocated to relevent vehicle groups.[10] The proportion of incremental costs so allocated will be the same as the proportion which remaining overhead costs bear to total fixed costs excluding those allocated incrementally to governments. The division of incremental costs between individual vehicles within the group should be either on an equal basis per standard vehicle unit or by the techniques of discriminatory pricing discussed below. In either case the appropriate contribution can be collected as part of the registration fee.

The same proportion of the excess cf average fixed costs on access and local roads over the average level on all roads is incremental to the traffic using these roads. This cannot be charged as part of the registration fee, since those vehicles are not distinguished by type, though where the level of such excess costs varies between counties it might be possible to charge different registration fees in different counties. Otherwise, such costs are best assessed against the benefits to the responsible vehicles by property taxes; the benefits from which they are paid being shifted from vehicle owners to property owners automatically.

The remaining costs are overheads and common to all vehicles. There is no valid theory as to the best way of allocating these. Their extent as a proportion of total costs is difficult to estimate in the absence of data. Not only will all marginal costs of wear and tear, that is, all variable cost, have been met already, but charges made to discourage traffic because of the costs it causes to other traffic at times of congestion will go, not to the other traffic inconvenienced, but as a contribution to overhead costs. The incremental part of overheads will also have been met. The remaining costs will therefore be considerably smaller than might appear at first sight. They will still exist, however, and be greater the better the highway system. Poor highways have low overheads and high marginal costs, the latter resulting in comparatively high revenues from variable taxes to meet comparatively low overhead costs. High-class modern highways have high overheads, and if they are adequate to carry peak traffic volumes the contribution to overheads from marginal cost pricing at times of congestion will be comparatively small. Good highways will therefore be the biggest problem.

Any attempt to apply equity to the problem of allocating remaining overhead resorts to a matter of opinion, and if there is scope for political decisions anywhere in the field of highway finance it is here. The following method is designed without reference to concepts of equity but with the object of minimizing the undesirable effects of various taxes in discouraging worth-while traffic. If the burden is spread evenly over all vehicle users by a flat registration fee per vehicle unit the marginal user will be discouraged; and the higher the fee the more traffic will be lost. If it is raised by additional fuel tax or tolls, marginal trips will be discouraged. To a certain extent, therefore, the choice is between discouraging the marginal user altogether and discouraging the marginal trip, neither of which is desirable. On the whole it is probable that more traffic would be discouraged in the form of marginal trips by all users than of marginal users altogether. An increased fuel tax will make users think about each trip, but a higher registration fee which raises the same total revenue, by being indispensable to the most valuable trips as well as the least important, will probably have comparatively little effect in reducing the total number of users. Each user derives a certain amount of total net benefit, over and above all costs borne by him, on those trips which he chooses to make at costs composed of vehicle and users' personal costs and those taxes so far described. This net benefit could be expressed in money terms as the maximum amount which could be taken from him without causing him to give up motoring. This is the maximum amount which could be collected from him in the form of a registration fee. While such a fee might result in his undertaking fewer trips, since the loss of purchasing power would force him to reduce his aggregate consumption and motoring might be one of the things affected, this income effect would probably be small. If the registration fee were at a level reasonably representing his responsibility for overhead costs this reduction in the number of trips would not be undesirable, since it would be caused by a reduction in the value to him of the marginal trip to a level below its marginal cost. The maximum amount which could be collected from him by a variable tax would be considerably less, since the imposition of a variable tax would increase the cost per trip, causing him to reduce the number of trips undertaken, and therefore reduce the amount of net benefit available for taxation. Furthermore, the imposition of more than a nominal variable tax would considerably reduce the number of trips undertaken, and this reduction would be undesirable as it would involve trips which might be worth more to him than their marginal cost, although less than marginal cost plus the additional variable tax. The advantage of the non-variable tax as a means of collecting overhead costs is therefore clear. A greater maximum contribution can be obtained from each person than by a variable tax, and there are no undesirable disincentive effects on the number of trips undertaken; whereas a variable tax would necessarily have undesirable disincentive effects.

If the variable or non-variable taxes which might be levied to recoup overhead costs would be at the same rates for all motorists, it is possible that the disincentive effect of the non-variable tax might be greater in some cases than that of the variable tax. A person undertaking a small number of trips would pay little in additional variable tax, and would reduce the number of his trips, whereas if he were charged a flat fee at the same level as a motorist undertaking a large number of trips he might give up motoring altogether. Any possible advantage of the variable tax arising from this ability to discriminate between motorists according to their mileage, is, however, offset by the possibility of using discriminatory non-variable taxes.

Vehicle taxes which discriminate among types of vehicles on bases other than costs are in practice quite common. The normal motives for discrimination have no economic validity, however, and must be rejected in devising an economic pricing system. One such basis, practised for some ears in the United Kingdom by differential rates of purchase tax on vehicles, is to levy higher rates of tax on luxury vehicles than on essential vehicles. Economic theory recognizes no distinction between necessities and luxuries, and there is certainly no justification for regarding commercial vehicles which will be used in the process of producing goods and services for consumers as essential, while regarding private passenger cars which yield services directly to consumers as luxuries. Concepts of necessity might play their part in the field of sumptuary taxation for general purposes, but they have no place in a pricing system. Similarly the ability-to-pay principle, which is often held to justify registration fees and purchase taxes which are progressive with the value of the vehicle, might be acceptable for purposes of sumptuary taxation but not as part of a pricing system. Motives of political expediency, which often account for (though are rarely admitted in defence of) lower taxation of farm trucks, are likewise completely irrelevant to a pricing system.

There is, however, one reason for discrimination in the allocation of overhead costs which is economically defensible. All traffic which is prepared to pay marginal costs is worth while, but if overhead costs are allocated on an equal basis per vehicle unit some of this worth-while traffic will be lost. If that traffic could be exempted there would be neither loss of revenue nor loss of worth-while traffic. This is illustrated in diagram 19 where *DD* is the demand curve to operate motor vehicle units as a function of the registration fee, on the assumption that all variable taxes will be paid. Curves I and II are rectangular

Diagram 19

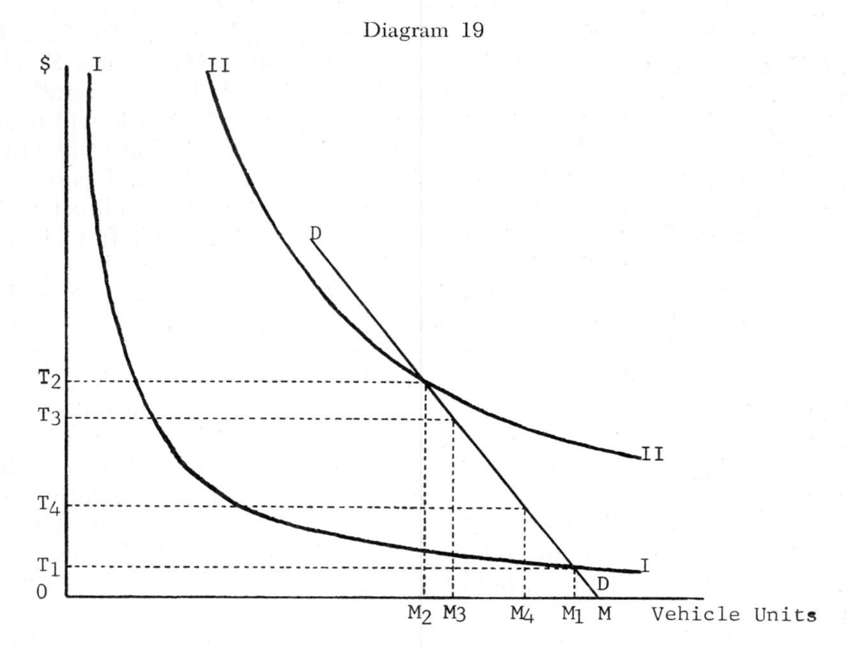

hyperbolas, the rectangular area under each being constant, and each representing a different assumed level of remaining overhead costs. If remaining overheads are at level I, a registration fee of OT_1 per vehicle unit would recoup these costs, but discourage M_1M worth-while traffic. If remaining overheads are at level II, however, a flat fee of OT_2 would be necessary, and this would discourage M_2M worth-while traffic. If the fee of OT_1 could be applied only to OM_1 traffic, M_1M being exempted, or OT_2 applied only to OM_2 and M_2M exempted, the revenues collected would be the same but no worth-while traffic would be lost. It can readily be seen that the higher the level of remaining overhead costs the greater is the amount of worth-while traffic which would be lost as a result of the imposition of a registration fee at a fixed amount per vehicle unit. Where remaining overheads are low the potential loss of traffic from a flat fee might be so small that it does not justify the additional complexity of discriminatory taxation, but where they are high some form of discrimination might be worth while.

The technique of discrimination is to charge a higher fee to those vehicles which constitute the left side of the demand curve, i.e., those which would derive the greatest net benefit. This is not in recognition of any benefit principle, however, and does not accept the argument that those deriving the greatest benefit should pay the most in highway taxes. There is no reason to determine tax rates in any way proportionally to benefits, but it is necessary to recognize that it is impossible to charge any vehicle more than the benefit which it will derive from highway use. Those vehicles paying the highest rates need pay no more than they would pay under a flat fee system, and if more than two rates are used it is possible for all vehicles to pay less than the flat fee rate. Thus while it would be necessary to charge a flat fee of OT_2 if remaining overhead costs are at level II, these costs could be recouped by a fee of OT_3 on OM_3 units, OT_4 on M_3M_4 units and nothing on the remaining M_4M units. Similarly, this system is not one of charging what the traffic will bear, but rather the realistic approach of not attempting to charge more than the traffic will bear.

The problem of using a discriminatory system in practice is that of identifying those vehicles which constitute the left-hand side of the demand curve. This problem can be approached either indirectly or directly. The indirect method is to use the fact that increases in property values which result from highway improvements are due to the shifting of vehicle benefits. Areas where property values rise most are these served by vehicles which derive the greatest net benefit. While these vehicles cannot be identified the benefits derived by them can be taxed after they have been shifted to property owners. Additional property taxes on those areas where property values have risen considerably as a result of highway improvement is therefore a simple and practical technique of recouping a contribution to overhead costs from those vehicles deriving extensive benefits from highway improvements, while imposing no burden on vehicles deriving small benefits, and therefore having no undesirable disincentive effect. It might be possible to derive from this source a sufficent contribution to overhead costs to reduce remaining overheads to a level at which the disincentive effect of a flat registration fee would be so small as to make further discrimination unnecessary.

The direct method of charging differential registration fees would involve identifying those vehicles on which higher rates could be charged. Where separate data on different types of vehicles are available from the estimation of demand for purposes of analysis, this might be accomplished without great difficulty. If heavy vehicles lie to the left of light vehicles in the demand curve of diagram 19 then discrimination by weight will achieve the desired result. If common carriers are to the left of private carriers, or buses to the left of trucks, then discrimination by type of service is indicated. Where no such data are available there is, of course, no basis for discrimination, and a flat fee per vehicle unit is then the only practicable system. Where discrimination is used the rates should be so arranged as to discourage as little worth-while traffic as possible, and at this minimum figure to discriminate as little as is necessary.

The only remaining forms of tax not so far discussed are such minor items as drivers' licence fees. These have no advantages over the registration fee as sources of highway revenue, but might be levied at low rates to cover the relevant administration costs of the regulatory authority.

CHAPTER TEN

THE ADMINISTRATIVE SYSTEM

THE PROVISION OF HIGHWAYS involves basically the same problems as any other economic activity. Scarce resources must be used to satisfy human wants by the provision of goods and services, and decisions must be made as to how much of our resources will be devoted to one particular service, and who is going to make the necessary sacrifice. Most economic activities fall conveniently into one of two classes: the first encompassing the production of goods and services that will be of benefit to individuals and will be paid for by individuals; the second encompassing those activities that will be of benefit to the community as a whole and will be paid for by the community as a whole. In our economic system decisions concerning the first class are made in markets and the costs paid in the form of prices; decisions regarding the second class are made by governments and costs paid in the form of taxes.

Highways are exceptional, though not unique, in that they combine the essential characteristics of both classes. The market system is inadequate for such a task while the government department is inappropriate. The best organization known to us for such a case is the public authority, established and controlled by government, but operating more on the lines of a large corporation, free from detailed political control.

In previous chapters we have seen how the diverse factors which are relevant to the making of decisions can be combined. We must now examine the organization and functioning of a suitable administrative system. The opinions and skills of various groups of people must be brought together and we must first examine what role each is to play.

Highways are built to serve the people, both individually and collectively. The individual highway user, be he private motorist or operator of a commercial fleet of trucks, decides where he wants to go and how much travel is worth. The various levels of government, representing the community at large, must make decisions as to those values which affect the community as a whole: the value of good highways to defence of the nation; the value to be placed on such aspects of location and design as the spoiling of beauty spots, clearance of slum areas; the value to be placed on human life and injury as part of the costs of accidents; and so on.

While individuals and governments perform the vital role of placing values on the services which highways yield and the non-material costs which they incur, they are not part of the highway authority itself. The first problem is therefore one of communicating such values from the persons who determine them to the authority which must act on them. Communications between the authority and the various levels of government are comparatively simple, through the normal channels of correspondence, memoranda, and committees.

Communications with individual highway users are made more complicated by the numbers involved. The values placed upon travel by individuals must be aggregated into the demand curves used in the analysis of previous chapters. This is the responsibility of the first group of specialists within the highway authority, who must combine some of the skills of the economist with some of the engineer. Existing patterns of movement are assessed by traffic surveys, while the elasticity of demand is assessed by comparison of existing analogous cases with different cost levels. Forecasts of demand in future years are based largely on income and population trends. One division of the highway authority would be responsible for estimating demand patterns for any proposed project and would accumulate the necessary data and experience.

A second division of the authority, composed almost entirely of engineers, would be responsible for estimating the cost curves for any proposal. The costs of construction create little difficulty, but the costs of vehicle operation and the influence of highway improvements on vehicle costs are matters on which much research remains to be done. Accident rates would be converted to money costs at values determined by politicians, while the task of converting time saving to money would fall to the economist.

These two divisions would thus convert the basic data on which decisions are based into the demand and cost curves, or their mathematical equivalents, to which the planning analysis can be applied. The planning division would be responsible for applying the planning analysis to the data. It would be composed of a small group of economists concerned with the technique of analysis, and a staff to undertake the necessary calculations, with the aid of computers.

A fourth division would be concerned with the financial problems of raising the necessary funds for implementation of worth-while projects. To this division would fall the task of devising an appropriate pricing system, arranging bond issues, and so on. The appropriate levies on gasoline and tires, registration fees, tolls and levies on property are essentially prices for highway use rather than taxes proper. Their imposition and enforcement, however, would in some cases have to be by co-operation with the appropriate legislative authority. Whether the highway authority would be given the authority to impose such levies, or whether they would have to be imposed and changed by the legislature, would not affect their basic nature. They are prices, not taxes, to be determined by the highway authority on the principles developed in previous chapters, and the proceeds from which would be the revenues of the highway authority. From these revenues the authority would meet the costs of building, maintaining, and administering the highway system, including the interest on and amortization of any bond issues made for highway financing.

In addition to these basic divisions of the highway authority concerned with planning, there would of course be the normal administrative departments organizing the construction and maintenance of the highway system.

The Structure of Highway Authorities

Almost all levels of governments in most countries have some responsibility for highways, and unless existing constitutions are to be radically changed the jurisdiction of each highway authority would coincide with that of the government which establishes it. The structure of authorities will therefore differ among particular countries. In the United States and Canada the major responsibility would fall on state or provincial authorities, with varying degrees of authority being delegated to county and municipal governments, while the role of the federal government authority would be more extensive in the United States than in Canada. In the United Kingdom the national authority would play a leading role, with local roads being administered by local authorities. No single structure will suit all countries, and since this book is not directed specifically at any one country the most that can be done is to lay down general principles.

The two essential elements of any structure of highway authorities are separation and co-operation. Since the territorial limits of jurisdiction of various levels of government overlap, the separation of responsibility for highways must be on some basis other than geography. The conventional method is by classifying highways into federal, state or provincial, and local roads, though the criteria of classification are by no means uniform or always clear. While some system of classification according to responsible authority is essential in any area where more than one authority has responsibility, it must be remembered that the highway network is a whole which cannot be divided into completely separated systems. A major improvement on a provincial highway, for example, can have significant repercussions on local and federal highways in the same area. Similar problems exist between geographically separated jurisdictions in the vicinity of the border. Provincial highways must clearly be related in terms of both location and capacity at the boundaries. If an integrated highway system is to result it is therefore essential on the one hand that the responsibilities of different authorities be defined, and on the other hand that a large measure of co-operation should exist among authorities at the planning stage.

These dual requirements can be met by combining an appropriate classification of highways with use of the planning analysis outlined in previous chapters. Since the use of the planning analysis involves specialist staff and extensive calculation it will normally be more economic for the smaller authorities to use the facilities of the larger authorities for this purpose. No loss of autonomy would be involved thereby, since the application of the planning analysis is essentially a process of detailed calculation rather than one of decision-making. Decisions as to the value to be placed on such aspects of the problem as slum clearance, and the preservation of beauty spots, which might be primarily the concern of a local government, would still be made at the appropriate level, even though the calculations based on such values would be undertaken at the provincial level.[1] The advantages of centralizing the application of the planning analysis are numerous. A large planning division could exploit the economies of computer techniques which become economical only on a large scale. The analysis would be uniform throughout the area, making co-operation between authorities more simple. A consistent body of data and experience would be built up in a central pool, and co-operation between provinces, and relations with federal authorities would be greatly simplified where each province has a competent planning staff. The federal authority might well be organized on a regional basis, and by working in close co-operation with provincial authorities as well as other regions of the federal authority, a framework for interprovincial co-operation would exist, while the technique used in the planning analysis would be uniform. The adoption of uniform techniques is of great value where different authorities are involved in the making of major decisions.

Under a unitary system of government such as exists in Great Britain the regional offices of a central authority would fulfil the functions of the state or provincial authorities under a federal system.

While the planning analysis might be thus centralized within each province, each authority would be responsible, in conjunction with its own government, in the making of value judgments affecting highways under its own jurisdiction. The administrative functions involved in the organization of construction and maintenance would also rest with each appropriate authority.

With a mutual spirit of co-operation the major problems of relations among authorities would exist not in the planning or administrative operations, but in the financing of highways. To these problems we must now turn our attention.

Financial Relations among Authorities

In chapter IX we developed the principles on which a pricing system for a single highway network should be based. The first stage was to assess the variable levies which would reflect the excess of marginal cost over average vehicle and users' personal costs. Perfection in this process is precluded by the practical impossibility of assessing different rates on each section of road, except in those few cases where it is worth levying tolls. It is therefore necessary in practice to assess the best compromise or average rate which must be applied on all roads in the network.

When we consider a hierarchy of highway authorities, the first stage is for each to assess the optimum rates of variable levies for the roads under its jurisdiction. It is immaterial whether local authorities undertake this task independently, or whether it is performed by the planning staff of the provincial authority on behalf of each local authority separately, since in either case the calculation is based on data yielded by the planning analysis. However, it is not possible to levy different rates on highways under different jurisdictions in the same area, nor is it practicable to have different rates for each local area. It is, however, possible to levy different rates for large areas. Adjacent states in the United States, or provinces in Canada, levy different rates of gasoline tax, but the one rate is normally uniform within the state or province. If we assume that the present system represents the limit of practical possibility, it will be necessary for the various authorities, local, provincial and federal, to agree on a single set of rates for each province. This can be most easily achieved by use of a weighted average, where the optimum rate for each authority with jurisdiction in the province is weighted by the total vehicle unit miles on highways under that authority's jurisdiction in the area. Variable levies on gasoline and tires would then be imposed throughout the province at these weighted average rates. The proceeds of such levies would be distributed among the various authorities in proportion to total vehicle unit mileage on highways under each authority's jurisdiction.

The second and third stages in building up the pricing system do not involve such close co-operation among authorities. Each authority can separately impose such tolls on roads under its jurisdiction as might be desirable, though where alternative routes lie on different road systems any tolls would have to bear an appropriate relationship to avoid the uneconomic diversion of traffic.

Similarly, each authority can claim against each level of government those incremental costs incurred for the provision of community benefits.

The bulk of remaining costs must be levied against the registration fee, partly on an incremental basis, and possibly incorporating some degree of discrimination, as outlined in chapter IX. The provincial and federal authorities would each devise an appropriate scale to cover its remaining costs. A similar scale would be computed to represent remaining costs of local authorities. The registration fee for each class of vehicle would be the sum of these and each authority would receive its appropriate share. While such a system involves no difficulties with respect to charging remaining costs of the provincial and federal authorities, the share of the local authorities is more complex, both in terms of the proportion of remaining costs collected in this way, and the distribution of the proceeds among authorities.

Different conditions apply in different areas and each province or state would have to devise its own best formula for this purpose. The simplest would be to aggregate remaining costs of local authorities; increase the provincial scale of registration fees by the proportion which remaining local authority costs bear to remaining provincial authority costs, and distribute the proceeds according to need. This might be considered inequitable in some cases, however, since it would involve levying on all traffic the remaining costs of local access roads. The following is suggested as an alternative system to overcome this objection, but involves one constant, the value of which would have to be determined by each province.

Remaining costs of the provincial authority would be divided by total vehicle unit miles on provincial highways to arrive at a standard cost per vehicle-unit-mile. This standard amount, multiplied by an appropriate constant, would be the amount to be collected by the registration fee as a contribution, per vehicle-unit-mile, to remaining costs on local highways. Each local authority would receive in total this standard contribution multiplied by the total vehicle-unit-miles on highways under its jurisdiction. The sum of these amounts would be raised by the third portion of the registration fee, but the distribution among vehicle types might well be different from that for the provincial or federal portion of the registration fee. If, for example, heavy trucks are prohibited from local highways, none of the total amount to be collected for local authorities from the registration fee would be assessed against this class of vehicle.

Since the object of such a formula is to avoid charging all traffic a share of the purely local costs of access roads, the amount received by local authorities in this way would fall short of their remaining costs. The deficiency is the responsibility of local traffic, and as was shown in chapter VIII this can most easily be collected as part of the property tax, since remaining local vehicle benefits are shifted to property owners. Each local authority would be responsible for collecting its remaining costs by appropriate levies on property.

While the above suggested system of financial relations among the various highway authorities might appear complicated at first reading, its application is comparatively simple. The necessary data on vehicle unit mileage on each highway network would be available from periodic traffic surveys which would be necessary anyway for use of the planning analysis. Once the data are known and the exact formulae determined, application of the system is achieved by the use of calculating machines. The resulting system of levies on gasoline and so on and the registration fee would be uniform throughout the province and paid as at present by the motorist.

Minor fees, such as those for drivers' licences, would be collected by the responsible authority to cover the administrative costs involved.

Taxation Proper

It has been argued throughout this book that levies on such items as gasoline and registration fees, which raise revenue specifically for the purpose of meeting highway costs, are essentially in the nature of a pricing system rather than taxation. The question now arises how far the highway transport industry should be subjected to taxation proper, that is, to the imposition of taxes to raise revenue for non-highway purposes. Highway history is fraught with examples of the use for other purposes of funds raised from taxes on highway use, often to such an extent that the construction and maintenance of an adequate highway system is jeopardized. This process has come to be known as diversion and has aroused considerable controversy.

It is often argued that diversion is necessarily a bad thing, and in its pure sense of using for one purpose revenues raised for another there are good reasons for condemning it. The term is more generally used, however, to include the use of funds raised from one sector of the economy for the benefit of another. This cannot be completely condemned since it is the essential process of any budgetary or fiscal policy. Few object to the use of revenue from tobacco or liquor taxes to pay for defence or education, though particular rates of tax or expenditures might be attacked, and there is no fundamental difference between taxing gasoline used for pleasure motoring and taxing tobacco to raise revenues for defence or education. It is not the use of gasoline tax revenues for non-highway purposes which is objectionable, but their misuse; yet the two are inseparable when the revenues are not raised for any specific purpose.

Where diversion has led to inadequate amounts being spent on highways two problems are involved, and pointless controversy rages because these are not kept distinct. On the one hand, the charges for highway services, raised for highway expenditures, are too low, and as a result too little is available for expenditure on highways; and on the other hand, pure taxes on the highway transport industry might be too high. While the lessening of diversion, that is, the reallocation of highway tax revenues from non-highway expenditures to highway expenditures would solve both of these problems if they are present to the same extent, it is by no means certain that they are present to the same extent. If they are not, then it means that either aggregate revenues from highway taxes are too high or too low. Controversy will rage indefinitely unless there is some agreed basis for determining the amount which should be raised for highway expenditures and the amount of pure taxation which should be imposed on highway users to raise revenue for non-highway purposes.

This book has been concerned with the rational determination of the optimum amount to spend on highways and the formulation of a pricing system. The extent to which pure taxation (for non-highway purposes) should be superimposed on this structure of charges for highway use is a separate problem.

Determination of the pricing system is an administrative function in just the same way that fixing railway freight rates or electricity charges is administrative. As such it is not an appropriate function of a minister of finance, provincial treasurer, or other politican. Pure taxation, or subsidy, is the function of the appropriate minister, and the fact that these two parts of highway "taxes" should be determined by different criteria, by different persons, and for different purposes, makes it vital that they be kept distinct, although they might well be collected at the same time and by the same means. Any transfers between the highway authority and the general treasury which are parts of our system of highway financing are not taxes or subsidies and should be treated accordingly. As we have seen, part of the costs of some highways might be paid by governments to cover the incremental costs of different plans adopted for reasons of defence or unemployment policy. Similarly if capital from the general treasury is used for highway building, interest and amortization would be paid from the "road fund" to the general treasury. These are payments for services and must not be confused with pure taxes on highways or vehicles. It is only if pricing and taxation are kept distinct that the former can be based on economic principles.

It does not follow that there should be no sumptuary taxation of highways or vehicles. Whether this is desirable, and if so to what extent, is a purely political question and economic theory can help no more than in any other case of the ethics of various taxes. We can say, however, that when it is decided to achieve certain aims by taxation, some methods of achieving them will be less undesirable than others from certain economic aspects. In order to achieve the optimum use of resources it is undesirable that the balance of economic competition be upset. Thus, if it is decided to raise a certain total revenue from taxation of industry, it is desirable to spread the burden evenly. Private passenger cars, for example, might well be subjected to the same rate of sales tax as other durable consumer goods. It is particularly important that any tax burden be imposed equally on competing forms of transport. Where highways are taxed more heavily than railways some traffic might be diverted to railways when it could be carried more economically by road. In the United States railways pay property tax on right of way, but highways do not. Dividends on railway stock are subject to income tax, but interest on toll-road bonds is not. Such discrimination can only encourage an uneconomic use of resources.

As with all taxation it is important that the form of the tax be chosen to match the objective of the tax. If the object is simply to raise revenue, then a registration tax (over and above the registration fee for highway services), or sales tax on vehicles, will probably have less distorting effects than a fuel tax. The ideal way to raise revenue would be to impose a lump-sum tax on the highway authority and let the authority recoup it from highway users as though it were part of overhead costs, since the objectives of the charging system will apply in both cases, that is, to raise the necessary revenue with the minimum undesirable consequences on traffic. If the objective is to reduce fuel consumption as an alternative to rationing, as for example in England during the Suez crisis, then clearly a fuel tax will be most appropriate. If it is decided to discriminate between types of traffic, for example, to tax private more than commercial traffic, this can be done easily with a registration tax or sales tax, and could be achieved with the fuel tax by use of coloured gasoline. Since pure tax rates are likely to vary more than highway charges, observation of the effects of changes might provide useful information on the elasticity of demand for road travel.

Thus it appears that there is no reason why highway transport, like other industries, should not bear its share of the burden of sumptuary taxation; but there is a very good reason why it should not bear a much heavier burden than other industries, while the highway authority loses its right to charge adequately for highway services in the process. If taxation were kept distinct from pricing, as it is in other industries, taxation would be clearly seen as a political process to which the normal principles of equity and fiscal policy apply, and about which the usual controversy over taxation would rage. But the pricing system would be based on purely economic principles, be largely free of controversy, and adequate revenue would be raised to finance such highway expenditures as are worth making (as determined by the planning analysis) without fear of diversion.

CHAPTER ELEVEN

CONCLUSIONS

THE OBJECT OF THIS BOOK has been to develop a rigorous framework of analysis, or technique of thinking, by which rational decisions can be made concerning the complex problems of planning an optimum highway system. From the planning analysis we proceeded to the financial problems and the criteria for formulating an optimum system of levies or charges for highway use. Finally, the administrative machinery by which such techniques of planning and financing could be implemented has been outlined.

To adopt all of the conclusions reached at once would call for substantial reorganization of our whole constitutional and administrative machinery. The impossibility of doing this in the practical world, particularly in a federal country where authority is shared by many governments, is obvious. This realistic limitation does not mean, however, that the whole theory is no more than an academic exercise in utopia with no relevance to the real world. New techniques and ideas are rarely adopted suddenly; they grow in acceptance as their value is proved in the practical world of experience. There is no reason why individual parts of this analysis should not be tried on a piecemeal basis, without the immediate acceptance of other parts. Individual highway authorities, or even individual officials within such authorities, might apply the techniques of analysis to specific problems. Experience gained in this way would then be disseminated through meetings and conferences of highway officials, and the theory, as modified if necessary in the light of practical experience, would gain wider acceptance. At the same time problems encountered in the evaluation of variables needed in the analysis would lead to valuable research to establish the actual data and relationships.

Acceptance and use of the planning analysis is the first stage. Implementation of the pricing analysis is more difficult, since gasoline taxes, registration fees, and so on, are at present determined by politicians. Highway officials are, however, in a position to advise ministers on such matters. Once a scientific basis of planning is in use a strong case can be made for such changes in existing tax rates as might be indicated. As governments become more convinced of the value of analysis as the basis of determining optimum tax rates they will be more and more inclined to accept the advice of officials of the highway department on such matters. Once this stage is reached the final transition to a public highway authority rather than a government department, with the power to determine its own pricing system, becomes possible. At the same time it becomes of little importance. For if governments and officials co-operate and keep clear in their own minds the distinction between prices and taxes, the right decisions will be made no matter who has the responsibility for making them.

While this book has been concerned specifically with highways, the technique of analysis is equally applicable to many other cases of public utilities. The variables involved might be different, the means of application in terms of a pricing system and administrative machinery would have to fit the particular case, but the basic process of relating numerous relevant factors into a logical framework of analysis on the basis of which rational decisions can be made is essentially the same. Experience gained in one field will be valuable in another, and as the application of economic analysis to such problems becomes more widespread it will be possible not only to make rational decisions in individual cases, but to compare on the same basis the relative claims of different public utilities on our scarce resources. Such a basis of comparison and consistent system of analysis will be of increasing importance for our economy in the years to come as the role of the public utility, to which the normal process of a free market cannot be applied, continues to expand.

APPENDIX

THE MARGINAL RULE IN WELFARE THEORY

THE BASIC CRITERION of planning used throughout this analysis is that marginal social cost should equal marginal social benefit as a necessary condition for welfare maximization. The use of such a rule in one part of an otherwise imperfect economy is open to serious objection, however. The purpose of this appendix is to examine the objections and attempt to justify our procedure. The problem of measurement of benefit as the area under the demand curve, and the necessary assumption that the distribution of income is equitable, have been examined in chapter IV. Our concern here is with problems of resource allocation.

Welfare economists have developed a series of marginal equivalencies which together constitute the necessary conditions for a Paretian optimum, defined as a position in which it would be impossible to make anyone better off without simultaneously making someone else worse off. There is general agreement today that the simultaneous fulfilment of all the marginal equivalencies throughout the economy necessarily brings about an optimum situation. In practice, however, this is little more than an exercise in utopia since it is impossible in the practical world to satisfy all the necessary conditions throughout the economy. This led to examination of the hypothesis that where it is not possible to satisfy all the necessary conditions it is best to satisfy as many as possible, and that the satisfaction of more conditions is a sufficient criterion for an increase in welfare. Unfortunately no such simple criterion exists. It has been shown[1] that, where it is not possible to satisfy one or more of the necessary conditions, nothing can in general be said about whether or not it is desirable to satisfy others. In fact, where one condition cannot be satisfied it might be best deliberately to violate other conditions, or the same conditions in other sectors of the economy. The "general theory of second best" enables us to calculate the optimum degree of fulfillment of some conditions, given restraints on the satisfaction of others, but its application calls for extensive data which are in practice never available. Welfare economics can therefore be said to have completed two stages of a three-stage progression towards practical usefulness.

The first stage was the discovery of the necessary and sufficient conditions for a Paretian optimum, which when combined with value judgments about the distribution of income and wealth determine that economic situation which would maximize welfare. To apply such conditions, however, would call for complete knowledge and complete power in a world where in practice both knowledge and power are severely limited.

The second stage was the discovery that a piecemeal approach to the problem might do harm rather than good; that if we do not have complete power the rules of what we should do with what power we have might be substantially different from the rules which should be followed in the same situation if we had complete power. This discovery, made independently in specific cases by different writers has been generalized by Lipsey and Lancaster as the general theory of second best. While this takes account of incomplete power by recognizing the impossibility of satisfying certain conditions, it does not help greatly in the practical world since its application still requires perfect knowledge, and can call for far more detailed knowledge than would be necessary for the application of the conditions for a Paretian optimum if we had complete power.

The third stage is the recognition of the fact that in the real world we have neither perfect power nor perfect knowledge, and that any practical application of welfare economics resorts to extracting what guidance we can from the knowledge at our disposal as to the best use of the power we have. This is necessarily an imprecise business since the best policy will vary with different possible values for the data about which we know very little, and often nothing at all. The attitude of most welfare economists when confronted with the complexity and imperfection of the real world is to argue that since we have insufficient knowledge it is impossible to prove conclusively that one policy is the best or even that it is better than any other policy and that therefore in general nothing can be said.

The present author, however, while agreeing that the current state of welfare theory is inadequate to prove conclusively that any one policy is the best in the circumstances, is firmly convinced that the theory has reached the stage where it can offer some guidance and lead to policies which are in general a considerable improvement on those that would be followed in the absence of any guidance at all. It is on the basis of this conviction that the present book has been written. The technique used has been to apply essentially Paretian conditions to the field of highway planning, since these are the only set of rules which it is possible to apply in practice. The general theory of second best shows that, since these same conditions are not applied throughout the economy, their application in one particular case does not necessarily lead to an optimum solution. We must now examine, therefore, what it is about the imperfections in the rest of the economy which might invalidate our conclusions, and how seriously wrong a policy based on a partial application of the Paretian conditions might actually be. These conditions have been applied to three problems: whether it is worth building a new highway and if so on what scale; the optimum level of use of a highway once built; and an optimum pricing system. These three applications must be examined in turn.

1. *Optimum Scale of New Construction*

The construction of a new highway is analogous to the building of any new industrial plant, and while various scales might be possible for a new plant each is in general indivisible. In the private sector of the economy new projects are undertaken when there is the expectation that total revenue will exceed total cost, while the criterion used above for highway projects is that total benefit should exceed total cost. Two differences exist between these criteria. In the absence of perfect price discrimination revenues can never be as high as benefits, and some consumers of the product are charged less than they would be prepared to pay, giving rise to consumers' surplus. A criterion based on revenue excludes consumers' surplus whereas a criterion based on benefit includes it.

Secondly, the private sector takes little if any account of external effects, whereas these are included in the highway planning analysis outlined above. Since consumers' surplus is excluded from the criterion used in the private sector it is possible that some private projects are discarded as unprofitable, while if they were assessed on a basis of total benefit they would be worth while, and might show a net benefit in excess of that which is held to justify a highway project of comparable magnitude. The use of the criteria advocated in this book would then result in resources being used for highway construction when they could have yielded a greater net benefit in some other use. This would, of course, be a sub-optimum allocation of resources, and would be a case where the market value of resources used underestimates the opportunity cost. In order to assess the seriousness of this objection to the use of our criterion, we must examine how widely revenues and benefits are likely to differ in the private enterprise sector of the economy from which the resources used in highway construction might be diverted. The resources are diverted mainly from the construction industry, with the exception of land which might be diverted either from use as construction sites, or from agricultural use. The resources could have been used for productive purposes in any of three market situations.

In a competitive situation, where the size of the individual production unit is small in relation to the total market, the use of either a revenue or a benefit criterion results in the same resource allocation. Additional units are undertaken so long as demand price is above construction cost, and the same number of units would be undertaken whether the criterion of revenue or benefit is applied. Indeed, since each unit is a marginal unit, revenue and benefit coincide. The consumers' surplus on intra-marginal units exists, of course, but is not significant at the planning stage where each additional unit is planned individually. Such competitive conditions exist in agriculture in the use of land which might be diverted to highway use, and in that part of the construction industry concerned with housing, apartment blocks, office buildings for rent, and the like. In so far as the resources used for highway construction come from this type of activity, therefore, the use of our criterion is not likely to result in any misallocation of resources.

At the other end of the scale of market situations lie those large indivisible construction projects such as railways and dams which might be undertaken by either private or public enterprise. In such cases in the private sector the difference between revenues, in the absence of price discrimination, and benefits can be considerable. However, it is in just these enterprises that price discrimination is common, and if perfect price discrimination is practised revenues and benefits coincide. In practice price discrimination is never perfect, but in the case of railways (a case with which we are particularly concerned since it is a competing transport medium) extensive use of price discrimination is common, and the divergence between revenues and benefits will therefore be considerably less than might at first be thought. Nevertheless, there will be some excess of benefit over revenue which might result in a proposed railway construction being rejected on a revenue criterion whereas it might have been held to be justified by the criterion of benefit advocated for highways. Where benefits in excess of possible revenues are held to justify a subsidy to the railway on a sufficient scale to bring about its construction, this offsets the misallocation of resources between highways and railways which might otherwise result. Since such subsidies from the public purse to aid railway construction are quite common, there is little reason to suppose that any extensive misallocation of resources between railways and highways would result from the use of the benefit criterion advocated above for highway planning. In the case of large single projects undertaken by the public sector the extent of possible resource misallocation between these and highways depends on the planning criterion adopted for such projects. It is of course advocated that the criteria developed in this book for highways be applied in a similar fashion to other public works projects, and if this is done no misallocation would result.

It is in the third possible type of market situation from which resources might be diverted for highway construction that the greatest danger of misallocation of resources arises. The construction of industrial plants by firms operating in markets of monopoly or oligopoly will be determined by the comparison of marginal cost and marginal revenue, that is by the amount which an additional plant would add to the total revenue of the firm in relation to the cost of such a plant. The addition to revenue might fall considerably below the addition to benefit, partly because the addition to total output will be large and price discrimination impossible, and partly because an increase in output resulting from expansion of plant might, by forcing a reduction in product price, reduce the revenue being derived from the output of existing capacity. The result is that many such projects which would be worth while if judged by the benefit criterion will not appear profitable, and too few of our resources will be used in this sector of the economy. This is of course the standard argument of underproduction in monopoly markets and the resulting misallocation of resources which occurs in a private enterprise economy characterized by varying degrees of imperfection in markets. It is not possible to devise a technique of highway planning which will redress the balance of resource allocation in the rest of the economy, and we are concerned with the problem only in so far as the danger exists that use of our criterion might make this situation worse. This would be possible only if highway construction results in the use of resources which would otherwise have been used for projects which would yield a greater net benefit. Where the resources are acquired in free markets, the only way in which other users would be persuaded to sacrifice them for highway purposes is by the increased demand for such resources so increasing their prices as to make the alternative uses no longer worth while. Since such price increases would apply to all uses of such resources, competitive as well as non-competitive, much of the required volume of resources would be released from alternative uses in competitive markets, and the resultant price rise would be less than would be necessary to procure their release entirely from imperfectly competitive alternative uses. Further, the competitive alternative uses will involve marginal projects from which resources are readily released by a price increase, with no undesirable loss of net benefit; whereas the projects in markets of monopoly or oligopoly are largely those on which pure profits would be anticipated, and these would be discarded to release resources only if the resource price increase raises costs by more than the anticipated pure profits. At the same time that the cost of building new plants is increased by competition from highways for construction materials, the anticipated costs of operating such plants are reduced by the savings in transport costs brought about by the highway improvement, either directly through reduced trucking costs, or indirectly through competitive railway freight rate reductions. The net effect of these influences on alternative uses of resources in monopolistic or oligopolistic markets is impossible to determine, but it would appear that they are to a certain extent counterbalancing. Any discrepancy remaining from the use of the more inclusive concept of benefit in highway planning, rather than the more restrictive concept of increased revenue used by private enterprise, is further offset by the difference in the treatment of costs.

Private enterprise generally takes account only of the direct costs of a project, whereas the criterion of cost advocated for highway planning includes external economies and diseconomies, largely in the form of community costs, negative and positive. The negative community costs of unemployment relief, slum clearance, and defence do not affect the comparison of criteria used for highways and other purposes. At a time when resources are unemployed, they can be used for highway construction without the necessity to divert them from other productive uses. Justification of resource use for slum clearance and defence must be compared with the criteria used for other comparable government expenditures, and the necessary correspondence is assured by having these negative costs assessed by the appropriate government departments and ministers. The inclusion of positive community costs, however, reflects a recognition of costs which would not normally be considered by private enterprise. Accordingly the criterion advocated for highway planning tends to result in higher estimates for both benefits and costs than would arise for revenues and costs by the criteria used by private enterprise in markets of monopoly or oligopoly. To the extent that both sides of the account are higher the net effect is unaltered.

In conclusion we can say that by comparison with alternative resource uses in competitive markets the criteria advocated for highway planning might result in estimates of benefit comparable with private concepts of revenue, but higher estimates of costs as a result of the inclusion of external effects. The result might be the use of too few resources for highways and too many in competitive industry. However, by comparison with alternative uses in markets of monopoly or oligopoly our criteria for highway planning might result in higher estimates of both benefits and costs, and accordingly the possibility that too many resources would be devoted to highways and too few to monopolistic industry. Since competitive and monopolistic industry are out of balance anyway it is impossible for highways to be in balance with both at the same time. If we had all the data necessary to apply the theory of second best we might find the cut-off point for highway projects to be where the ratio of benefits to costs is slightly greater than, or slightly less than, one, and it might even be different for different projects. But we do not have such data, and some criterion is desperately needed. It is the author's contention that the use of a cut-off ratio of one is as good an approximation to the optimum as it is possible to achieve in practice. Its use would ensure consistency within the field of highway planning, and while the absence of data makes it impossible to prove that it would result in the best allocation of resources between highways and other uses, there is no way of knowing, nor even any reason to believe, that it would result in overinvestment in highways rather than underinvestment, or *vice versa*.

2. *Optimum Level of Highway Use*

The criterion used in the above analysis for the optimum volume of traffic on a given highway, is the intersection of the demand curve with the marginal cost curve. This implies that a trip is worth making if the user considers it worth more than the marginal cost involved. In other sectors of the economy, however, a good is consumed only if the consumer considers it worth more than its price and price might exceed marginal cost, especially under monopoly conditions if marginal revenue equals marginal cost and price exceeds marginal revenue. It could therefore be argued that if price generally exceeds marginal cost in the private sector of the economy, the criterion for an optimum volume of traffic should similarly involve an excess of demand price over marginal cost, since otherwise resources would be used on traffic to yield a return just equal to cost where they might have been used for some other purpose to yield a return of benefit in excess of cost. This misallocation of resources could, of course, occur only when resources are used which would otherwise have been used for some other purpose.

The items comprising marginal cost in our analysis are entirely variable costs; the variable highway costs of maintenance and depreciation, vehicle operating costs, users' personal costs, and community costs. The highway costs involve the use of construction materials in much the same way for maintenance as for construction of highways. The arguments used above to justify the marginal criterion for construction therefore apply equally to maintenance. Vehicle operating costs are composed of the prices paid for vehicles, fuel, tires, repairs, and so on. These are reckoned at market price rather than marginal cost of their production, and in so far as there is an excess of price over marginal cost in the rest of the economy on these items, account will automatically be taken of it in our analysis; for if demand price for the trip equals marginal cost of the trip, and marginal cost of the trip includes the prices of the components of vehicle operating costs, then to the extent that marginal cost of the trip consists of vehicle operating costs, the demand price for the trip must bear the same ratio to the marginal cost of producing the components of vehicle operating costs that their market prices bear to marginal cost.

Users' personal costs consist of time, inconvenience, and risk of accident. Inconvenience and risk are hardly allocable resources, but time is. The valuation of time at the wage rate would underestimate the opportunity cost of time only if time spent travelling would otherwise be used in productive activities where the value of the marginal product of labour exceeds the wage rate. If the time involved would otherwise be devoted to leisure then no misallocation results on the assumption that the marginal value of leisure is the wage rate. The former condition applies mainly to the time of persons employed to travel, and the criterion for such employment is that the marginal revenue product of labour should equal the wage rate. Where the marginal revenue product of employed persons travelling is less than the value of marginal product, it is the former which determines the demand price of the trip. If the ratio of the marginal revenue product of labour to the value of marginal product in the case of employed persons travelling is the same as that in the alternative productive use of such labour then no misallocation of labour would result. The same conclusion holds with regard to the wage rate and the marginal resource cost of labour, where the labour of employed persons travelling is purchased under conditions of monopsony. To the extent that persons might devote leisure time to travel, which time would otherwise be used in productive employment where the value of marginal product exceeds the wage rate, then there is some case for arguing that the equivalence of demand price to marginal cost would result in the allocation of too much labour to travel. The extent of this misallocation would appear to be small, however, and its correction would involve valuing time spent travelling at more than the wage rate. Since the customary practice in highway planning is to value time at considerably less than the wage rate, this serves only to show that the valuation of time advocated in our analysis is a marked improvement on current practice.

In conclusion, we can say that there is little ground for the argument that our criterion of equalizing the demand price of a trip with marginal cost would result in misallocation of resources between travel and other productive uses. As in the case of determining the optimum scale of new construction there might be some tendency for too many construction materials to be devoted to highway maintenance and too few to other uses in markets of imperfect competition, and there might be some use of leisure time in travelling which would otherwise be put to productive use yielding greater benefit, though the existence of this tendency is questionable and its extent certainly small. Thus, while it is possible that if complete data were available we should find that our criterion tends to err on one side or the other, in the absence of enough data to apply the theory of second best the criterion advocated in this analysis would appear to be as good a rule as it is possible to find.

3. *Optimum Pricing System*

The use of the marginal cost pricing approach to the financing of highways follows from the criterion for the optimum traffic volume. If we accept that the optimum volume is that at which demand price equals marginal cost, then the pricing system should be designed to achieve that volume. The volume actually travelling will be that at which demand price equals average cost to the user including taxes used to recoup highway costs. The pricing system must therefore be so arranged that average cost to the user equals marginal cost of the trip in order that the actual volume will be the optimum. Since the user automatically bears the average operating costs (vehicle costs and users' personal costs) the appropriate variable tax must be equal to the excess of marginal cost of the trip over average operating cost.

The taxation system used to recoup total costs in excess of those met directly by the user or covered by the variable tax is more difficult. On the assumption that the social marginal utility of money is roughly the same for all persons, which follows from the assumption of an equitable income distribution, and providing that the system used does not lead to a significant deviation from optimum traffic volumes, then the criteria for a supplementary tax structure resort to a matter of welfare distribution rather than welfare maximization. This is entirely a matter of value judgments and the structure advocated above reflects the judgments of the author. There is no way of proving that these are either better or worse than any alternative set of value judgments which satisfy the necessary conditions.

Conclusion

The best economic policy, in the sense of that leading to a Paretian optimum, can be achieved by satisfying a defined set of marginal equivalencies. This is impossible in practice since it would call for complete power and perfect knowledge of which we have neither. The second best policy might involve violating the Paretian conditions and satisfying others which are determined by the general theory of second best. This allows for the fact that in practice power is incomplete, but is still not a guide to practical policy for it still necessitates perfect knowledge. The third best policy, which allows for the constraints of both incomplete power and imperfect knowledge, is at once the most vague and the most useful, for it is the best that we can do in practice. This book attempts to define the third best policy for highway planning. The conclusion reached is that while conditions of the second best policy might differ from those of the best, the third best resorts to a partial application of the conditions of a Paretian optimum. This conclusion is based on an examination of the particular conditions of this case; it is not a general rule for guidance of practical policy in all applications of partial welfare analysis.

NOTES

CHAPTER TWO

1. See, for example, *Informational Report by Committee on Planning and Design Policies on Road User Benefit Analysis for Highway Improvements* (Washington, D.C.: American Association of State Highway Officials), 1955, Preface, p. (a). "The method of analysis applies only for alternates of location and design of the same general highway facility . . . Analysis is made by a comparison of the relations of annual road user savings to the annual capital costs for the logical alternates in location and design that have the same over-all traffic movements on the highway or connecting highways affected . . .
"A benefit ratio so determined is not suitable for use directly in comparison of highway projects that have dissimilar traffic, terrain, and design conditions. It is not suitable for priority determination of projects on an area or State-wide basis."
2. See, however, the Appendix for a more detailed examination of this point.
3. The new price is, of course, the marginal opportunity cost of resource use, which might exceed the average opportunity cost.

CHAPTER FOUR

1. The reader interested in the intricacies of this problem is advised to consult the following: Jules Dupuit, *De l'utilité et de sa mesure* (Turin, 1934); Alfred Marshall, *Principles of Economics* (8th ed., London, 1952), bk. 3, chap. vi; J. R. Hicks, *A Revision of Demand Theory* (Oxford, 1956).
2. Net benefit is the excess of total benefit over cost. The area V_1FCE is composed of a reduction in cost of V_1FGE on OX_1 units, with no change in total benefit; plus FCG which is the excess of total benefit (X_1FCX) over total cost (X_1GCX) on the additional X_1X units demanded at the lower price.
3. See Antony Terry, "Road Plans by Electronics," *Sunday Times,* London, Aug. 9, 1959: "By cutting out . . . the 'arithmetical drudgery' of road planning . . . costly calculation which took up the time of hundreds of skilled engineers for months, can now be done in a few days."
4. *Economic Survey of Europe, 1956,* United Nations.

CHAPTER FIVE

1. For Californian experience of this see David R. Levin, "Problems of Highway Right-of-Way and Control of Access," *Better Roads,* Dec., 1955.
2. These examples are taken from J. Labatut and W. J. Lane, eds., *Highways in our National Life, chap.* xxvii, "Permanence of Right-of-Way," by David R. Levin (Princeton, 1950).
3. In an economic sense "development costs" once incurred are not true costs. Once a tunnel is bored it will never need doing again, and repayments on the initial borrowed capital cease to be true costs of the highway. They remain financial commitments, however. Throughout this thesis the terms "costs," "total costs," and "average total costs" are used in a general sense to include all financial expenditure and obligations. Types of costs are identified by adjectives, e.g., "marginal costs," "variable costs," "historic costs." The extent to which historic or development costs should be regarded as true costs is discussed in chap. vii.
4. See *Highway Practice in the U.S.A.,* Public Roads Administration (Washington, D.C., 1949), p. 79.
5. See n. 7.
6. These figures also illustrate the misleading conclusions drawn from an inappropriate use of the distinction between urban and rural operation. Practical capacity is defined as, "the maximum number of vehicles that can pass a given point on a lane or roadway during one hour under the prevailing roadway and traffic conditions, without unreasonable delay or restriction in the drivers' freedom to maneuver." The reason given for the lower practical capacity of the rural motorway is that speeds are higher. In no case is a road more isolated from its surroundings than in that of a limited access motorway, however; and given a road of the same design characteristics, its capacity at a given speed will be the same whether it

is surrounded by fields or skyscrapers. The only reason for the difference in practical capacity therefore is that a greater restriction on freedom of speed and manoeuvrability is considered "reasonable" in the case of urban than in the case of rural motorways. It is not because speeds are lower that the urban motorway carries more traffic; it is because there is more traffic that speeds are lower.

7. See *Informational Report by Committee on Planning and Design Policies on Road User Benefit Analysis for Highway Improvements,* American Association of State Highway Officials (Washington, D.C., 1955).

8. *Ibid.,* p. 5.

9. R. F. Newby, "The Cost of Delay at Traffic Signals in Central London," Department of Scientific and Industrial Research, Road Research Laboratory, Nov. 1955.

10. For experience of this in Los Angeles see Lloyd Aldrich, "Freeway System Benefits" (Los Angeles, 1954).

11. A.A.S.H.O., *Informational Report,* p. 112.

12. J. H. Jones, *Road Accidents,* Report submitted to the Minister of Transport (London, 1946).

13. *Ibid.*

14. *Ibid.*

15. *Ibid.*

16. *Ibid.*

17. D. J. Reynolds, "The Cost of Road Accidents," *Journal of the Royal Statistical Society,* vol. 119, part IV, 1956.

18. For information on these see J. T. Duff and D. A. de C. Bellamy, "Surveys to Determine the Origin and Destination of Traffic," Department of Scientific and Industrial Research, Road Research Laboratory, 1955; *Origin and Destination Surveys: Methods and Costs,* Highway Research Board, Washington, D.C., Bulletin no. 76.

19. See Max-Erich Feuchtinger "Some aspects of the design and planning of urban motorways," B.R.F., Urban Motorways Conference, 1956.

20. See G. K. Zipf, "The $(P_1P_2)/D$ Hypothesis: On the Intercity Movement of Persons," American Sociological Review, vol. 11, no. 6, Dec., 1946, p. 677.

CHAPTER SIX

1. The best exposition of this, together with a geometric combination of a solvency quotient and a benefit quotient, is in C. A. McCullough and J. Beakey, *The Economics of Highway Planning,* Oregon State Highway Department, Technical Bulletin no. 7, 1938.

2. See Levin, "Problems of Highway Right-of-Way and Control of Access."

CHAPTER SEVEN

1. For the development of this distinction, and definitions, see chap. II.

2. See W. Rees Jeffreys, "How may new modern roads be financed?" *Road International,* no. 2., Spring, 1951.

CHAPTER EIGHT

1. For a fuller discussion of this see chap. I.

2. References on principles of taxation are listed in the appendix to this chapter.

3. The supply curve is not perfectly inelastic because although the amount of land is fixed, there are other uses for it besides building sites.

4. The latter assumption was made by the British Conference on Road and Rail Transport in 1932.

5. See *Motor Vehicle Taxation in Illinois,* Department of Public Works and Buildings, Division of Highways (Springfield, 1952).

6. These figures are typical of conditions in the U.S.A.

7. See, for example, M. Earl Campbell, "Considerations in the Assignment of Financial Responsibility," 1950.

8. C. B. Breed, *Report Upon Costs of Roads Required for Heavy Motor Vehicles Compared with Costs of Roads Adequate for Passenger Automobiles and Light Trucks,* Associated Railroads of Pennsylvania, 1933.

9. H. Tucker and M. C. Leager, *Highway Economics* (Scranton, Pa., 1942), chap. 13.

10. The "Martin Report" of the Interim Committee appointed by the Governor of the State of Oregon (Salem, 1937).

11. "Economic competition" is any competition which is economic, that is, which results in resources yielding the greatest benefit, and services being obtained at the lowest cost. Maintaining economic competition does not mean arranging to have different forms of transport offering the same service at the same price irrespective of cost because competition is considered to be "a good thing."

12. When average cost is falling, marginal cost must be less than average cost. Average cost, however, includes some items of community cost for which no compensation is paid. Marginal cost in excess of average vehicle and users' personal costs might therefore exceed average cost actually borne by the highway authority.

13. See, however, the appendix for a discussion of other objections.

14. The standard vehicle unit was defined in chap. II.

CHAPTER NINE

1. We are concerned with pricing of highway services, not taxation proper. Terms such as "gasoline tax" are so well sanctioned by usage, however, that to employ new terms could only lead to confusion. Throughout this chapter therefore, "taxes" will refer to parts of the pricing system. Taxation proper is considered in the next chapter.

2. R. G. Hennes, *Allocation of Road and Street Costs, part I: An Equitable Solution to the Problem,* Washington State Council for Highway Research, Seattle, 1956.

3. In the absence of any technical data, this assumption is as reasonable as any.

4. For fuller details and arguments on this tax see M. Earl Weller, *The Highway Use (Weight-Distance) Tax in the State of Oregon* (Salem, 1950); R. H. Baldock, "The Case for the Weight Mile Tax," *Road International,* no. 10, autumn, 1953; Yule Fisher and Barbara Bruce, *Weighed — and Found Wanting: The Experience of Eleven States Which Have Tried and Rejected The Ton Mile Tax and Similar Taxes,* National Highway Users' Conference (Washington, D.C., 1956); D. E. Brisbine and Yule Fisher, *The Ton Mile and Related 'Third Structure' Taxes,* N.H.U.C. (Washington, D.C., 1950).

5. See Brisbine and Fisher, *The Ton Mile and Related 'Third Structure' Taxes.*

6. See H. E. Davis, R. A. Moyer, N. Kennedy and H. S. Lapin, *Toll Road Developments and their Significance in the Provision of Expressways,* Institute of Transportation and Traffic Engineering, University of California, Research Report no. 11 (Berkeley, Calif., 1953) Table II, p. 31.

7. Georgia, Iowa, Kansas, Kentucky, Minnesota, New Mexico, Oklahoma, Tennessee, Utah, West Virginia, Wisconsin, and Idaho.

8. See Parsons, Brinckerhoff, Hall and Macdonald, *Report on the Traffic and Earnings of Ohio Turnpike Project, no. 1,* New York, May, 1952.

9. The standard vehicle unit as defined in chap. II.

10. See the discussion of the charging of remaining overheads in chap. VIII.

CHAPTER TEN

1. While the Canadian term "province" is used in this chapter, all that is said of the province and provincial authority would apply equally to the state and state authority in the United States.

APPENDIX

1. R. G. Lipsey and Kelvin Lancaster, "The General Theory of Second Best," *Review of Economic Studies,* vol. XXIV (1), 1956-57, pp. 11-32.

INDEX